A HOSPITABLE UNIVERSE

ADDRESSING ETHICAL AND SPIRITUAL CONCERNS IN LIGHT OF RECENT SCIENTIFIC DISCOVERIES

By
Rodolfo Gambini
with the collaboration of
Jorge Pullin

Imprint-academic.com

Published in the UK by
Imprint Academic, PO Box 200, Exeter EX5 5YX, UK

Distributed in the USA by
Ingram Book Company,
One Ingram Blvd., La Vergne, TN 37086, USA

ISBN 9781845409647

A CIP catalogue record for this book is available from the
British Library and US Library of Congress

Contents

Preface

The advances in the natural sciences of the last century have reconfigured our vision of the world. Of particular importance were the changes produced in 20th century physics: special and general relativity and quantum mechanics and more recently cosmology on one hand and biology, in particular the recent developments in genetics —in particular genomics— on the other.

This book attends to a growing need to reexamine various pessimist outlooks resulting from outdated ways of understanding the world. In particular, forms of nihilism that arose as criticisms of the traditional beliefs based on the discoveries of the natural sciences, mainly physics and biology of the 18th and 19th century in light of the mentioned recent advances.

The approach we choose is the following: we first present an introduction to the developments in physics. This introduction is essential if one intends to understand in detail the concepts that serve as a basis for the ontology elaborated in the text. The latter is based in an analysis of the basic objects of the quantum formalism. As is well known, quantum mechanics does not offer at the moment univocal answers to many questions of metaphysical character. Due to this we will choose a series of interpretations of quantum mechanics that admit an ontology of events and states that appears particularly suited to reshape the traditional philosophical systems based on classical physics.

In the second part we lay the basis of the quantum ontology and we observe that it leads to a renewed vision of the notion of emergence in complex systems and of top-down causation of the wholes on its constituent parts. The core of the third part of the book consists in extracting certain consequences of developments in biology and cosmology that highlight the centrality of life and the existence of a cosmos

9

more hospitable to life than previously thought. In the fourth part we focus on the development of a new form of natural religiousness.

The general spirit of the presentation is to lay out the problems from a vision where the scientific takes precedence over the philosophical. But also admitting that some form of philosophical reflection is key if one wishes to recover a unified vision of the world where human beings are part of a bigger unity and manifest in exceptional ways characteristics that are already present in the rest of the Universe.

Our intention is to foster a dialogue and to introduce the topics so expert philosophers could eventually propose more rigorous treatments or alternative approaches. The book is self-contained in that it introduces the basic physical notions required for the discussion. Since some of the issues in question are rather sophisticated, it requires a little bit of detail in the description of the physics, in particular the use of some elementary equations.

Montevideo, March 25th 2017

Acknowledgements

This work was supported in part by the National Science Foundation, the John Templeton Foundation, the Fundamental Questions Institute (FQXi), Agencia Nacional de Investigación e Innovación (ANII), Programa de Desarrollo de las Ciencias Basicas (PEDECIBA), the Hearne Institute for Theoretical Physics and the Center for Computation and Technology at the Louisiana State University.

Chapter 1

Introduction

"The breeze is cool and the sky blue. I love this life with abandon and wish to speak of it boldly: it makes me proud of my human condition. Yet people have often told me: there is nothing to be proud of. Yes there is: this sun, this sea my heart leaping with youth the salt taste of my body and the vast landscape in which tenderness and glory merge in blue and yellow" Albert Camus in 'Nuptials at Tipasa'.

We have all shared feelings of this kind towards nature and the Universe. We travel thousands of miles to visit places where these experiences are more powerful. They are not restricted to a particular kind of sensitivity, of the religious person, the scientist or the artist. They leave powerfully rooted memories in our spirit and they flame our love for life. According to our sensitivity, we will be more impacted by a sunny day or contemplating the starry sky in a summer night.

The meaning of these experiences changes from person to person. Camus was an existentialist, and he shared with them the belief in the "death of God" and of the human experience in a meaningless world. He recognized the intensity of his experiences without attributing them any further meaning that the one that arose in the moment, without pre-conceiving or conceptualizing it. These kinds of feelings are accessible to all men, but their interpretation depends on our upbringing and our natural tendencies and they lead us to react in different ways. In scientists like Carl Sagan, they left a feeling of admiration for the Universe that manifests repeatedly in their work. People like him are religious men exalted by a religion without God, motivated by the love of nature.

Others react by interpreting those experiences like the works of a God responsible for the Creation. In his article "Wonder and Skepticism" (1995), Sagan describes how that intense admiration for the night sky led him to science. He remarks that there does not exist science without such admiration, but at the same time he recognizes that there is no scientific quest without skepticism: "I was a child in a time of hope. I grew up when the expectations for science were very high: in the thirties and forties. I went to college in the early fifties, got my Ph.D. in 1960. There was a sense of optimism about science and the future. ...there was one aspect of that environment that, for some reason, struck me as different, and that was the stars. So I asked my friends what they were. They said, 'They're lights in the sky, kid.' [But] What were they? [Finally] My mother said to me, 'Look, we've just got you a library card..., get out a book and find the answer.'... The answer was that the Sun was a star, except very far away. The stars were suns; if you were close to them, they would look just like our sun. I tried to imagine how far away from the Sun you'd have to be for it to be as dim as a star. Of course I didn't know the inverse square law of light propagation; I hadn't a ghost of a chance of figuring it out. But it was clear to me that you'd have to be very far away. Farther away, probably, than New Jersey. The dazzling idea of a universe vast beyond imagining swept over me. It has stayed with me ever since. I sensed awe. And later on (it took me several years to find this), I realized that we were on a planet —a little, non-self-luminous world going around our star. And so all those other stars might have planets going around them. If planets, then life, intelligence, other Brooklyns.– who knew? The diversity of those possible worlds struck me...[Some years later I discovered that it was possible to become a professional scientist, that] you could spend all your time learning about the universe. It was a glorious day. [But soon I discovered that] Science involves a seemingly self-contradictory mix of attitudes: On the one hand it requires an almost complete openness to all ideas, no matter how bizarre and weird they sound, a propensity to wonder. As I walk along, my time slows down; I shrink in the direction of motion, and I get more massive. That's crazy! On the scale of the very small, the molecule can be in this position, in that position, but it is prohibited from being in any intermediate position. That's wild! But the first is a statement of special relativity, and the second is a consequence of quantum mechanics. Like it or not, that's the way the world is. If you insist that it's ridiculous, you will be forever closed to the major findings of science. But at the same time, science requires the most vigorous and uncompromising skepticism, because the vast majority of ideas are simply wrong, and the only way you can distinguish the right from the wrong, the wheat from the chaff, is by critical experiment and analysis. Too much openness and you accept every notion, idea, and hypothesis —which is tantamount to knowing nothing. Too much skepticism— especially rejection of new ideas before they are adequately tested —and you're not only unpleasantly grumpy, but also closed to the advance of science. A judicious mix is what we need."

The religious person, or at least the theist, feels the same admiration as the scientist and in the same sense mistrusts it, not in order to moderate it with a certain dose of skepticism, but to go beyond it as a distraction in the admiration of God. A well known passage in St. Augustine's Confessions (Burke 2012) says: "Too late loved I You, O You Beauty of ancient days, yet ever new! too late I loved you! And behold, You were within, and I abroad, and there I searched for you; deformed I, plunging amid those fair forms which You had made. You were with me, but I was not with You. Things held me far from You, which, unless they were in You, were not at all. You called, and shouted, and burst my deafness. You flashed, shone, and scattered my blindness. You breathed odours, and I drew in breath and panted for You. I tasted, and hunger and thirst. You touched me, and I burned for Your peace."

Thus, the believer in a personal God represses her feelings of admiration for nature in search of a superior reality that is super-natural and the scientist attempts to go beyond it in search of a superior order: the discovery of the laws of nature that rule the Universe.

This questioning of the emotions that the wonders of the material Universe offers us is not limited to the believer or the scientist, it is shared by everyone. Only the children seem to enjoy fully and without reservations these experiences. When life faces us with worries or setbacks we begin to believe that such magnificence is alien to us and sometimes even hostile. It seems to obey its own strange rules and they do not seem to be related to our own necessities, desires or worries.

In order to live with such opposing feelings we have sought different kinds of answers. It can be said that every culture, every mythology has produced one. In the West for a long time the theist answer was the dominant one: there exists, beyond the Universe a personal God that has created it. A God that guarantees in one way or another that in the end "everything will be alright". For some, the few, that God restores our harmony with the Universe. St. Francis of Assisi is the paradigm of this possibility. For the majority the division between nature and a transcendent God totally disconnected from it is the predominant view, one can only endure the misfortunes of the world with faith and the hope in a better future "in the other world".

But the role played by the formal religious systems in the Western World has suffered a noticeable decadence in the last few centuries. There are several causes for the phenomenon, and in some cases they predate the birth of modernity, being in many examples associated to specific details of the Judeo-Christian religious tradition.

For the medieval man, in fear of a world in which the guarantees of a relatively safe and peaceful life had disappeared, living in constant fear of tomorrow, at the mercy of the arbitrariness of the feudal lords or the attacks of foreign enemies, religion was more than a theological system. It was the psychological matrix around which the entire life was structured from birth to death. The images, symbols and rituals were the immediate experience and the kernel of all certainties. The central role of the Christian world, as was inherited from ancient times, would gradually disappear in a prolonged process that starts in the Renaissance, leaving man to drift in a world whose objectivity had been stripped of meaning. The feeling of loss only manifested itself progressively. Initially, the secularization process was perceived as the exaltation of the trust of humankind in its infinite possibilities of creation and domination of the world. The increasingly valued new freedoms acquired since the Renaissance are one of the distinctive characteristics of modernity.

However, the progressive disintegration of the globalizing vision of Christianity, left a blank slate in entire spheres of human sensitivity, the trust in justice and history as well as the spiritual forces that determine our behavior and are the basis of all ethical behavior, were left temporarily without support. Gradually, the immense vacuum left by the "death of God" invaded the conscience of men. At the end of the 19th and beginning of the 20th Century a series of of secular movements attempt to, in a growing and more or less conscious and systematic way, to fill the void with great intellectual constructions. They serve, whatever their content of truth, as "substitute creeds" or "mythologies" that provide a framework for the world. Marxism and Psychoanalysis are two of the most notable examples of such constructions.

The three distinctive elements of the modern period, the development of scientific rationality, the birth of the project of personal and social autonomy due to Illuminism and the rational and systematic organization of economic processes due to Capitalism are the determining factors of the religious decay of the times. They are part of a dual process of transformation, of gains and losses and ultimately of maturity of the human being.

Of the three, the development of science, in particular of mathematical physics, is the main trigger of the secularization process mentioned above. The mathematical, and therefore exact, description of the physical world is an innovation of incalculable consequences and a singular example of the duality between gains and losses that we referred to. It offers the possibility of knowing exactly the world and leads to the discovery of a cosmos subject to strict and universal rules that describe, for instance, with the same principles the motions of planets and other celestial bodies and those of everyday objects. At the same time it strips objects of characteristics related to the human senses like color, taste or smell, they are considered irrelevant at the time of studying their mechanical behavior. The progressive abstraction in the perception of the world is, among many other factors associated with the birth of the natural sciences, what has led to the Illuminist vision of the world. This in turn contributed to the development of the great democratic revolution, the modern critical thinking and a secular vision of knowledge as a instrument of human progress which has been at the roots of the intellectual and cultural tradition of the West.

This modern conception of the world faces men and women or our time with the contradiction between a beautiful yet immense complex and strange Universe and the personal and human sphere and not leaving much margin for consolation. It basically consists in recognizing the otherness of the Universe and to build a sphere of the human, confined and finite that focuses on the possibilities that current life offers and at most hopes for a better future for humankind in which we are excluded by our finite lifetimes. Bertrand Russell (1925/2004) bravely describes the alternative of humankind that cannot believe anymore in the supernatural in the current world: "I believe that when I die I shall rot, and nothing of my ego will survive. I am not young and I love life. But I should scorn to shiver with terror at the thought of annihilation. Happiness is nonetheless true happiness because it must come to an end, nor do thought and love lose their value because they are not everlasting. Many a man has borne himself proudly on the scaffold; surely the same pride should teach us to think truly about man's place in the world.

Even if the open windows of science at first make us shiver after the cozy indoor warmth of traditional humanizing myths, in the end the fresh air brings vigor, and the great spaces have a splendour of their own."

Although as Richard Dawkins (2003) observes there may exist certain greatness in this "bleak and cold" vision of life we cannot but ask ourselves time and time again if it may not be possible within the strict confines of the natural to reintegrate humankind to a universe that appears detached from us. We think that a possible way opens if life and in particular the human being endowed with a personality and consciousness can be reinserted in the universe as a central element who has a protagonic role in its development. The search for a third way is today only possible in a strictly naturalist framework. Will the development of science that brought us to this point be able to provide a vision beyond what for many is a dichotomy between illusion and desperation?

The road ahead forces us to make explicit the basic hypotheses, to understand some of the advances of physics and biology and their implications. We do not pretend more than showing that current science opens other possibilities that future investigations may make more explicit. We attempt to explore what can be considered a minimalist position about religion without pretending to question who go beyond in their beliefs and convictions. We will see that the mere existence of such a possibility offers a glimmer of hope of constructing a vision of a human existence which is not divorced from nature.

Chapter 2

Naturalism, physicalism, emergence

2.1 Naturalism

The influence of science has only grown in the last three centuries. Although we recognize today that we can never have absolute certainties, science seems to be the main instrument to give us reliable beliefs about the world we live in. We daily add knowledge about the universe, life and ourselves. We understand in an increasingly better way the behavior of stars or the way in which living organisms transmit their traits to their descendants. Many mysteries about the world have been revealed and it may only be a matter of time to uncover others.

Nature, as described by science, seems to progressively form part of a consistent system of laws whose consequences account for everything that happens without any intervention of an external agent. That has led to raise the naturalist methodological hypothesis to a metaphysical one that establishes that "reality is exhausted by nature, containing nothing 'supernatural', and that the scientific method should be used to investigate all areas of reality, including the 'human spirit' " (Papineau 2015). It is clearly not a precise definition and it may be understood in more than one way. That the scientific method should be used does not mean necessarily that it is enough to analyze any sphere of reality. Take for instance mental phenomena or artistic expressions, that they all in-

volve some physical substrates does not mean that the latter exhaust
their contents. For instance, we will not discuss the existence of a sphere
of pure feelings (which we will later call "phenomenal"); for instance,
pain, scents, or a certain hue of blue. What is generally assumed is
that such a sphere has a neuro-physiological counterpart. In fact, many
philosophers adopt a naturalist view of the mental processes to explain
how they affect our actions and as a consequence have manifest physical effects. If the naturalism that we adopt here must account for the
totality of human activities, including the rational, artistic and social
ones it faces very high hurdles that prima facie appear very difficult to
satisfy. In particular to explain the origin of human creativity or its
personal ethical convictions. The notion of material world has evolved
through time. If the physical vision of the world were restricted to the
mere motion of elementary corpuscles there would be little room left for
life and even less to understand the human spirit and its deeds. We will
see that recent advances of the physical and life sciences lead to a conception of matter and nature so rich that make possible satisfying such
high expectations. It will therefore be necessary to familiarize ourselves
with the new advances of science if we wish to understand them and
reflect on their consequences.

It is clear that if one starts from the hypotheses that in order to have
the greatest degree of certitude in our conclusions we must limit ourselves to adopt the naturalist vision that stems from science, we need to
reexamine our religious options in this context. It seems like a long and
difficult journey whose answers are not apparent nor immediate, but the
reward may be to know that our beliefs have a certain degree of scientific support and, as we shall see, to discover that we live in a Universe
that is more hospitable and compatible with human expectations.

This tough road full of difficulties is the one chosen by science and
it constitutes, as Nietzsche observed with great irony and perspicacity
in "On the genealogy of morals" (1887/2009) the latest and most noble
form of asceticism, since in spite of their deliberate skepticism, scientists
still are believers since they "still have faith in truth".

2.2 Physicalism

Since we are physicists, we will follow a long and extended tradition
among those who practice our profession known as physicalism. Some

kind of physicalist position is implicitly or explicitly shared by a large number of physicists. Einstein (1918) explained his view in this way: "The supreme test of the physicist is to arrive at those universal elementary laws from which the cosmos can be built up by pure deduction." In his Scientific American article Stephen Weinberg (1974) reformulated this goal with the following words "One of man's enduring hopes has been to find a few simple general laws that would explain why nature with all its seeming complexity and variety is the way it is. At the present moment the closest we can come to a unified view of nature is a description in terms of elementary particles and their mutual interactions."

Both Weinberg and Einstein, like the majority of physicists, admit without questioning that there exist propositions and statements that are true about nature and that at some point will become part of our physical theories. The existence of those statements would ultimately confirm the approximate success of our physical theories, that are still provisional and being developed today. We will call physical laws those true statements towards our physical theories progress. They are the "universal elementary laws" of Einstein or the "few simple general laws that would explain why nature... is the way it is" of Weinberg. Most of our current physical laws would therefore not be "physical laws" in that sense. Many reflections about the underlying reality of the world begin with the hypothesis that such "physical laws" exist and that our current laws reflect them approximately. Einstein and Weinberg also admit some kind of epistemological reductionism: the behaviors observed at various levels of reality can be deduced from the fundamental physical laws, it nevertheless can be more convenient to describe them through other sciences like chemistry or biology.

Although we share the fundamental aspects of this position it is worthwhile to analyze it in detail and not limit ourselves to oversimplifying proposals. Sometimes scientificists fall into and that hold that all metaphysical or ontological considerations must be eliminated. On the contrary, we consider fundamental for any attempt to base a religious attitude of life on a naturalist view of the world to be guided by the desire "to understand how things in the broadest possible sense of the term hang together in the broadest possible sense of the term" (Sellars 1960/1963) and this is the task of philosophy.

But what is the true nature of physical laws? Two positions have emerged in the literature, the regularist and the necessitarian. The regularist position is that they derive their truthfulness from the actual connections between events and states of the world and therefore express only what does occur. The necessitarian, as Norman Swartz (2003/1985) claims "physical laws determine which connection can and cannot occur", therefore determining what must occur in certain circumstances. In our view, many philosophers who claim to be physicalist are necessitarian in their views and this has constrained their possibilities of elaborating coherent world views. To avoid those constraints, we will adopt a position that we call regularist physicalism. In this section we will discuss this notion in some detail.

Let us recall that for regularists, the physical laws are statements about regularities observed in events and states of the world. They are therefore descriptions about what happens in the world. Physical reality may transcend these laws that only describe its observed regularities. On the other hand, for the necessitarian "It is the physical laws that set the bounds on what can and cannot occur. For the Necessitarian, there is some sense in which physical laws have primacy over mere occurrences. Laws impose 'constraints' on what the world is and what occurs therein" (Swartz 2003/1985).

In physics, a modern way to hold a necessitarian position is that of Tegmark (2008), who claims that our universe could satisfy the following hypothesis, called Mathematical Universe Hypothesis (MUH): "Our external physical reality is a mathematical structure." He distinguishes between "the outside view or bird perspective of a mathematician studying the mathematical structure and the inside view or frog perspective of an observer living in it." This would be an extreme case of necessitarianism where the formal structure —the mathematical laws— precede and ultimately, determine, given a mathematical structure, how to compute the inside view from the outside view. This radical physicalist vision leads to claim that "even the languages, the notions and the common cultural heritage that we have evolved is dismissed as 'baggage', stripped of any fundamental status for describing the ultimate reality."

On the other hand, Einstein appears to have held a regularist point of view. "Being is always something which is mentally constructed by us, that is, something which we freely posit (in the logical sense). The justification of such constructs does not lie in their derivation from what is given by the senses. Such a type of derivation (in the sense of logical deducibility) is nowhere to be had, not even in the domain of pre-scientific thinking. The justification of the constructs, which represent reality for us, lies alone in their quality of making intelligible what is sensorily given..." Schilpp (1998). On other instance he adds "If two different peoples pursue physics independently of one another, they will create systems that certainly agree as regards the impressions ('elements' in Mach's sense). The mental constructions that the two devise for connecting these 'elements' can be vastly different" (Einstein 1917). Theories are therefore under-determined by the "impressions", although in practice physicists are unaware of this under-determination. It is because ours is not the situation of "two different peoples pursuing physics independently of one another." We are members of a common scientific community. Don Howard (2004) adds "Part of what it means to be a member of a such a community is that we have been taught to make our theoretical choices in accord with criteria or values that we hold in common." Summarizing, for a necessitarian laws have the primacy, in this case, one should think about the world in mathematical or logical terms, for a regularist the ultimate nature of the world is not determined by the physical laws that only describe the regularities that the processes of this world present. In the last case one could consider that the fundamental concepts used in the mathematical formalism of our physical theories represent elements of a reality whose nature transcends the theory and that we can only access directly from inside in Tegmark terms, that is from our total experience. In some sense we propose a view that inverts the hierarchical relation proposed by Tegmark, who incurs in what Whitehead calls "The fallacy of Misplaced concreteness". For example, according to Whitehead the notion of point is the result of a limiting process, he says: "among the primary elements of nature as apprehended in our immediate experience, there is no element whatever which possesses this character of simple location. ...[Instead,] I hold that by a process of constructive abstraction we can arrive at abstractions which are the simply located bits of material, and at other abstractions which are the minds included in the scientific scheme." (1925/1997).

2.3 Epistemological reductionism and ontological emergence

We take a physicalist view towards reduction inspired in the ideas of Einstein and Weinberg described above. Therefore, we will consider that reducing a given sphere of reality to another means to have the ability to deduce the behaviors of a given theory in terms of a deeper and more unifying one. As pointed out by Nagel (1979) this requires adding bridge principles linking both theories. These allow to connect the language of the theory being reduced to that of the more fundamental theory. In most cases the reduction implies a modification of the reduced theory. As an example, the wave theory of light modifies geometric optics, in which one treats light via rays. The latter theory is useful for analyzing certain simple phenomena in optics like the propagation of light through lenses or its reflection in mirrors. It fails, however, to explain interference and diffraction. In several examples strict reduction happens only as a limit. For instance, general relativity reproduces Newtons theory of gravity in the limit in which the speed of light is infinite and gravitational fields are weak. The reduction process goes hand in hand with the acknowledgment that the more fundamental theory is better than the reduced one, both in the range of described phenomena and in the accuracy of their description. In the mentioned example of wave optics that can describe diffraction whereas geometric optics cannot, the latter can be seen as a limit of vanishing wavelength of the former. With vanishing wavelength there are no waves and diffraction phenomena are as a consequence absent.

Physicalism is reductionist as it assumes that all laws of nature can be deduced from those of physics. This should not be confused with assuming that physics explains everything in nature. To start, there exist many phenomena, particularly in modern physics that cannot be explained. The decay of an energy level to another of an atom through the emission of a photon at a given time is not "explained" by quantum mechanics. The latter only makes statistical predictions and, more strongly, it emphatically denies the existence of any ulterior physical explanation of a given event. Regularist physicalism therefore does not claim that physical theories exhaustively describe all natural phenomena. It only states that regularities observed in science concerning the various levels of reality (e.g. chemistry, biology or psychology) are deducible from the lower level theories. These types of explanations do not deny the emergence of new behaviors and properties, what is being claimed is that reality is structured hierarchically.

When we speak of deducibility, we generally refer to a process that can be carried out in principle, even though it may not ever be carried out in practice. For example, although it is believed that molecular behavior can be explained in terms of quantum mechanics, the computational complexity in deducing the behavior of complex molecules like DNA from Schrödinger's equation prevents us from doing so explicitly (although given the rapid increase in computer power it might be in some future).

Regularities in phenomena are described by the laws of nature. Reduction is just the acknowledgment of the possibility of the existence of such phenomena in a world ruled by the laws of physics. As Wimsatt (2007) put it "Emergent phenomena... are often subject to surprising and revealing reductionistic explanations. But reductionists often misunderstand the consequences of their explanatory successes. Giving such explanations does not deny their importance or make them any less emergent-quite the contrary: it explains why and how they are important, and ineliminably so."

Basically, the reductionist vision in the sense mentioned above contributes to understanding the genesis of the most complex levels of reality in terms of the simpler ones. This kind of reductionism is usually known as epistemological since it refers to how our knowledge of a certain level, say for instance chemistry, can be deduced from a more fundamental theory, for example quantum mechanics.

Following Butterfield (2011), if we denote the fundamental or "bottom" theory Tb and by Tt the reduced ("top") theory, one would say that Tt is reducible to Tb if one can prove that Tt is logically deducible from Tb making use of some suitably chosen definitions. Let us consider the following example: accounting for the chemical bonds of a simple molecule of ionic hydrogen, formed by two atoms of hydrogen and removing one electron. It is relatively straightforward to show that a hydrogen atom and a proton will tend to form a molecule. One computes quantum mechanically the minimum energy that the system has when the electron orbits around one of the proton and the minimum energy when it orbits around both. One can see that the latter is lower than the former (Cohen et al. 1978). Due to this, after the molecule is formed it would require adding energy (or exerting a force) to separate it. That is how chemical bonds form. So this is an example of how the molecular structure of matter is explained in terms of quantum mechanics. Different molecules will require different studies, but in all cases where the calculation can be carried out in practice, the bonded configuration will be the one with the lowest energy. In complicated molecules there can be several configurations that correspond to local

minima of the energy function. All of them will be possible configurations of the molecule. The chemical bonds can originate in different
processes, depending on the molecule in question, but they can all be
explained in physical terms.

Reducing a phenomenon to another provides an explanation of it in
terms of the base theory, it does not replace the concept by another.
This kind of reductionism, which we can call epistemological, can be —
as we shall see—, compatible with the emergence of complex structures
whose properties are not determined by those of its parts. We will see
that quantum physics has a characteristic mechanism to produce the
emergence of wholes richer than their parts. The combination of simple
elements, for example to form a water molecule, produces a whole where
the component elements are altered to the point that they lose in part
the identity that they had when they were isolated.

The appearance of complex structures with emergent properties —of
wholes that are more than the sum of their parts— is a typical phenomenon and not an exceptional one, that repeats at different levels
and originates in an underlying property of quantum systems known as
entanglement. In entangled systems, for instance in a molecule, the complete system dictates the behavior of its components which are modified
and lose the identity they had as isolated components.

As we mentioned, both Einstein and Weinberg are epistemological
reductionists and hope that some day physics will allow to deduce completely, starting from its laws, the behavior of systems like the chemical
ones. Nevertheless, this is not contradictory with the emergence of complex entities —of wholes— richer than their parts. This phenomenon is,
as we will see, very well justified and leads us to believe in the so called
ontological emergence. Epistemological reductionism explains how the
universe is capable of evolutionary creating more complex entities and
ontological emergence (also known as strong emergence) implies that
systems can have qualities not directly traceable to the systems components. The new qualities are therefore ontologically irreducible. In
what follows we will discuss this form of emergence and how it may
appear physically in the quantum theory. The emergentist vision is key
in the religious naturalism and determines a strong contrast between
the conception of the Universe as a created object in a single act, and
a Universe where life and consciousness develop in a creative process
that can take different forms in different places. The contingent nature of this process that nevertheless appears capable of producing life

and consciousness beyond the details of the history of each place or each inhabitable planet is one of the most fascinating features of this naturalistic approach.

2.4   Religious naturalism

Religious naturalism considers that in order to determine if the most profound religious interests can be satisfied without recourse to additional postulates, we should base ourselves in the natural phenomena and processes. In general the religious naturalist understands religion in a broad sense, without restricting to interests and experiences of institutional religious groups but rather with the existential, spiritual and social concerns of all people. For the religious naturalist all human beings have experiences and concerns of religious nature. According to Wildman we understand that "religious naturalism rules out both supernaturalism and all views of ultimate reality as an active, focally aware entity, but... it affirms an ultimate reality in the axiological, (that is valuational) depth structures and dynamics of nature." The main purpose of this book is to explore what can be said of this "ultimate reality" and how it responds to our ethical, existential and religious concerns. In order to look for answers for this question we will proceed reviewing the results of science, in particular attempting to describe and interpret the fundamental physical theories. Also we will stress the key role of life in a universe whose properties seem to be finned tuned for its existence. With this starting point, we propose to develop an ontology and to explain the emergence of complex systems with new properties and causal capabilities not reducible to those of their parts. This characteristic of many complex systems, practically omnipresent in the quantum world is known as downward causation. It implies in particular that we cannot speak about isolated individuals without considering their environment in the broadest sense. The picture that emerges is one of a profoundly unitary reality.

Part I

A quick tour of
contemporary physics

Chapter 3

Birth and zenith of mechanism

3.1 Introduction

When one is preparing to attack a problem of any sort: practical, scientific or philosophical, one cannot get rid of preconceived ideas and prejudices. One cannot isolate them: they are part of what is usually called common sense. We react to problems based on certain knowledge resulting from our experiences, beliefs or previous studies. This knowledge is in many cases vague and sometimes contradictory. The difficulties one encounters when trying to understand the philosophical implications of current physical theories may come from inadequacies of common sense, too ingrained to be easily recognizable as erroneous.

The first great transformation of common sense due to the incorporation of a scientific theory was mainly due to the work of Galileo and Newton. They disentangled the problem of motion, which for thousands of years had not had a satisfactory solution. Through this, they set the scientific basis of physics. Galileo established some of the fundamental principles of the new science of motion, now called classical mechanics, and at the same time proposed the basis for its interpretation. The motions we perceive, be it a stone thrown in the air or a planet moving in the sky, are fairly complicated. So the analysis of motion begins with simpler cases. An observation that can serve as starting point is that objects at rest remain at rest unless an influence is exerted on them that puts them in motion.

Galileo noted that the predominant belief at the time, that objects free of influences tend to be at rest, is false. He realized that if an object is launched in a horizontal plane, it would continue moving longer and longer the more we reduce friction forces. Thus a coin sliding on a frozen lake will move farther than if we just try to slide it a sand beach. In fact Galileo observes that the coin ends up stopping due to the existence of external influences that act on it, the ones due to friction with the lake or the sand of the beach. Galileo's observation, a cornerstone of the new physics is that if no forces are exerted on a body, the latter moves uniformly, that is, with constant velocity in a straight line. This is called inertia law. From the same experiments Aristotle and Galileo reach different conclusions. For the former the conclusion was bodies that tend to rest, for the latter that bodies tend to maintain their speed. To understand how one goes from one conclusion to the other we need to take into account the evolution of thought between the Greek times and Galileo's times. He was convinced of the validity of the heliocentric hypothesis of Copernicus, to which he himself had contributed astronomical evidence. According to that hypothesis, the Earth has a daily motion of intrinsic rotation and, in addition, it revolves around the Sun together with its satellite, the Moon. Such hypothesis had been dismissed as absurd with arguments based on Aristotles physics and the predominant common sense at the time. The notable Danish astronomer Tycho Brahe argued against heliocentrism with the following argument recounted by Galileo himself: "According to the argument which convinced Tycho, and which is used against the motion of the earth in Galileo's own Trattato della Sfera, observation shows that 'heavy bodies... falling down from on high, go by a straight and vertical line to the surface of the earth. This is considered an irrefutable argument for the earth being motionless. For, if it made the diurnal rotation, a tower from whose top a rock was let fall, being carried by the whirling of the earth, would travel many hundreds of yards to the east in the time the rock would consume in its fall, and the rock ought to strike the earth that distance away from the base of the tower."' (Feyerabend 2010). Note that the argument has two parts. In the first one, it acknowledges an observed fact. In the second, the fact is analyzed using the principles of Aristotelian physics and a contradiction with observation is reached. From there it is concluded that the hypothesis of Earth rotation is false. To rebuke the argument,

Galileo accepts facts: it is indisputable that heavy bodies dropped from a tower fall in a straight line, but he will question the conceptual framework used to analyze those facts. The first idea he is going to criticize is that of the absolute character of motion implicit in Aristotelian physics. According to this point of view, it was thought that if a body like the Earth would move at great speed, it was absurd that such motion was not perceived by us. For Aristotle an object free of forces tends to rest. If the Earth rotated, the object would stop rotating with the Earth as soon as dropped from the tower. As a consequence, from the Earth it would be seen left behind the tower as it falls and its motion would describe an ample curve. With Aristotelian physics, only if the Earth does not rotate one would see the object fall vertically. Note that this is an argument that, in the perspective of the time, was very strong and was accepted a priori without even being incorporated explicitly in the principles of Aristotelian physics. Galileo begins by reminding us, following the Socratic method, that real motion is not always perceived. In other words, movement is relative and depends with respect to which system it is measured.

"Salviati: Now imagine yourself in a boat with your eyes fixed on a point of the sail yard. Do you think that because the boat is moving along briskly, you will have to move your eyes in order to keep your vision always on that point of the sail yard and to follow its motion? Simplicio: I am sure that I should not need to make any change at all; not just as to my vision, but if I had aimed a musket I should never have to move it a hairs-breadth to keep it aimed, no matter how the boat moved. Salviati: And this comes about because the motion of the ship confers upon the sail yard, it confers also upon you and upon your eyes, so that you need not move them a bit in order to gaze at the top of the sail yard...". Note that from Galileo's analysis one ends up with a critical view of the experiment that discovers that our perceptions are on many occasions tinted by non analyzed beliefs. Those who, facing Galileo's arguments stated "but I see that the Earth does not move" would not have understood this essential point. On it rests all of modern science.

Having established that we only perceive relative motions, what remains to be explained is why the stone falls next to the tower and why it is not left behind by it. Indeed, one of the explicit dynamical hypothesis of Aristotles physics is that objects on which no interaction acts tend to be at rest. Unless Aristotelian thought is modified, an object released from the top of a tower would start to lose speed and would be left behind with respect to the tower that is rotating with the Earth. With that aim, Galileo is led to introduce the concept of inertia. Recall that he had concluded that a body free of forces will continue to move indefinitely. The new concept of inertia, that establishes that a body free of forces will move with constant velocity, completes Galileo's refutation of the tower argument. Once released, the rock, acted upon only by gravity, will keep the component of its velocity parallel to the Earth unchanged due to its inertia, and as a consequence will not drift away from the tower. Summarizing, Galileo succeeds in proposing an interpretation in which internal coherence between the heliocentric hypothesis and the new principles of mechanics is manifest and where observed facts result naturally of the new laws. For that purpose he had to modify the primitive concept of motion, introducing the idea of relative motion, and to transform the concept of inertia. The following century will see the blossoming and of the interpretation of mechanics initiated by Galileo and its enormous impact not only on the development of physics but on a new world vision.

Newton would elevate mechanics to a totally dominant position in the thought of the next century by showing that the motion of planets can be deduced from simple and universal laws. Planetary motion results from two assumptions: one, of general character, establishes the connection between forces and changes of speed; the second establishes the particular form of the force between gravitating bodies. The same description stemming from these two assumptions explains the motion of a stone falling in the vicinity of the Earth and the motion of the Moon in its orbit. For Newton the world is composed of material particles. The state of motion of the system is, at a given instant, characterized by the positions and velocities of the particles and the forces that act upon them. From them, the laws of motion of Newton allow to determine the state of motion in any other instant. This is the basis of the mechanist/determinist vision of the world that would dominate scientific thoughts for centuries to come. The science of mechanics achieved great accomplishments in all its branches, in particular remarkable developments in astronomy and its ulterior application in many other areas apparently different and non-mechanical in a first glimpse, like hydrodynamics or kinetic theory of gases. All this led to the belief that it was possible to describe all natural phenomena in terms of simple forces between inalterable objects. Until the mid 19th century this will be the dominant vision. Helmholtz for instance claimed that "the problem of physical material science... is... to refer natural phenomena back to unchangeable attractive and repulsive forces whose intensity depends wholly upon distance. The solubility of this problem is the condition of the complete comprehensibility of nature". (Helmholtz 1853). It is this extremely poor vision of the resulting physical world of Newtonian physics, which reduces it to particles in motion that interact with predetermined forces, which led to conceive a Universe that is totally inhospitable and detached from human concerns.

3.2 The impact of classical mechanics on the thought of the 17th and 18th centuries

In order to illustrate how the new conceptual framework resulting from the Galilean–Newtonian new science of mechanics was applied and illuminated areas of knowledge that transcend physics, we will mention two in which the impact was deep and lasting: cosmology and Cartesian mechanism. Concerning the first example, it was inevitable that the vision of the Universe would change when it was discovered that celestial bodies did not belong in a sphere of infinite perfection as was conceived by the Greeks in particular by Aristotle. On the contrary, they obeyed the same universal laws that ruled the motion of everyday terrestrial objects. Such change of paradigm is at the basis of modern cosmology. Concerning the second example, it can be said that the impact was even more profound although not as positive since it led to an oversimplified version of natural phenomena.

As the French mathematician Pierre-Simon de Laplace claimed "Given for one instant an intelligence which could comprehend all the forces by which nature is animated and the respective situations of the beings who compose it...for [this intelligence] nothing would be uncertain and the future as the past would be present to its eyes" (Laplace 1814/2010). In this way nothing is left to the designs of such intelligence, only devoted to contemplate the inevitable development of events. Mechanist determinism pushed in that way to its last consequences left the world without any possible opening to novelty. In spite of the fact that Newton perceived science as a way of worshiping God, and considered his own work as a hymn to the divine power and intelligence, its effects, in particular the mechanist determinism left very little room for freedom and human responsibilities, and barely allows anything beyond a world of automatons. Voltaire commented that "it would be strange indeed, that all nature, all the planets, should obey eternal laws, and that should be a little animal, five feet high, who, in contempt of these laws, could act as he pleased, solely according to his caprice." (Damper 1968). In fact, as Laplace points out, in a mechanist Universe once the positions and velocities of the particles that compose it at a given instance and the forces they interact with are completely determined, there does not exist any possibility that an external action of God or a free action of a human may alter its development without violating the laws of Newton's mechanics. This property of the mechanism is usually referred to as "causal closure".

We will devote some attention to Cartesian mechanism since, in addition to illustrating the impact of classical mechanics on the philosophical thought of its time it has had consequences on common sense that persist till our days: it has complicated the understanding of the implications of 20th century physics and even today impacts negatively on how to conceive nature.

3.3   The birth of modern cosmology

In spite of its many critics, the Aristotelian vision of a finite and limited universe with the Earth at rest in its center and the stars occupying the outermost sphere rotating with eternal and perfect circular motion, was totally consistent with his physics. Indeed, as we have discussed, the latter did not admit an Earth in motion. Although the concept of an infinite universe was introduced first by Giordano Bruno before the birth of classical mechanics with Galileo and Newton, it is the work of the latter that turns into completely unacceptable the Aristotelian cosmology.

Indeed, the concepts of relativity and inertia exclude any possibility of identifying any point that is at absolute rest and could be the center of the universe. In classical mechanics all inertial systems, associated with particles free of forces and moving at constant speeds, are equivalent. It is not possible to distinguish in an absolute way a particle moving with constant velocity and one at rest. It is always possible to take the opposite point of view and observe the system in such a way that the particle that supposedly was in motion is now at rest, and reciprocally, the one that was at rest acquires a constant velocity.

On the other hand, classical mechanics describes motion in terms of changes of position of particles that evolve in a space purely geometric given by Euclidean geometry. Aristotle, who was not a geometer, conceived space as having a physical nature and not a purely geometrical one. In this he resembled more Einstein than Galileo or Newton. Indeed, for Einstein space is subject to dynamical laws and interacts with other material objects. That is the reason Aristotle can speak of a space limited by the celestial sphere. If he had been asked what was there outside the sphere he would have replied nothing there is no space nor no place outside. All space is in the interior. Einstein also speaks about a space that is like the surface of a sphere, therefore finite but unlimited and that can curve without having to think that there is something in the interior of the sphere. On the other hand both Galileo and Newton conceived space in Platonic purely geometric terms. Such space was unaltered by the presence of matter. It was the pure and infinite space of Euclidean geometry. To speak of a celestial sphere that limits the region where matter is distributed is totally unnatural with this point of view and it immediately leads to the question of what is outside the sphere.

The new physics crushes the celestial sphere. Stars, now considered as other Suns, are situated at enormous distances. In 1725 James Bradley would initiate a prolonged scientific effort to measure by trigonometric means, that is by measuring parallaxes, the distances to stars. Unfortunately, the lack of precision of instruments of the time prevented him from measuring the distances and only a century later could the first distances to stars be measured. Finally Kant, almost 70 years after Newton wrote his Principia, and inspired in the work of astronomer Thomas Wright, and even earlier observations of Galileo, lays down the hypothesis that our our Solar system belongs to an enormous planar system of stars, the Milky Way. Extrapolating, he proposes for the first time the idea of a universe composed of galaxies.

3.4 Cartesian mechanism

The new physics being created in the 17th century makes evident the contradictions of the Aristotelian thinking that had been at the core of medieval scholastics. The great task that thinkers of that century set out to execute is to elaborate a new philosophy consistent with the world vision provided by the new physics. That would in particular be the goal of Descartes' philosophy, who establishes explicitly the central role of physics: "The whole of philosophy is like a tree. The roots are metaphysics, the trunk is physics, and the other brunches emerging from the trunk are all the other sciences" (1644/2006). He also takes from the work of Galileo the notions that will lead him to define matter in terms of the mathematical concept of extension.

The foundations of the mechanist posture were established with remarkable clarity by Galileo long before the deterministic dynamical laws were laid out by Newton. Such posture states that the basic concepts needed to understand the idea of matter are those of number, space and time and establishes that "the book of the universe is written in the language of mathematics". It is him who first suggests a distinction between what would be later called the primary intrinsic properties of an object, among which are the concepts of space and time we just mentioned, and the secondary ones associated to the sensations of color, smell, taste or sound that designate changing states of our mind. He says "let us mentally suppress the living beings and their organs and these qualities disappear from the world."

This is the essence of the mechanist position that strips reality from any attribute that is not mathematical or at least mathematizable by the science of motion. Alfred North Whitehead describes this position eloquently when he states that "Thus nature gets credit which should in truth be reserved for ourselves: the rose for its scent, the nightingale for his song, and the sun for his radiance. The poets are entirely mistaken. They should address their lyrics to themselves, and should turn them into odes of self-congratulation on the excellency of human mind. Nature is a dull affair, soundless, scentless, colorless, only the hurrying of material, endlessly, meaninglessly."

The doctrine of two substances of Descartes is the natural continuation of the Galilean line of thought. Descartes starts by applying the principles of the new physics to the most general possible setting. Not only the planets and the inanimate objects will obey the strict deterministic laws of mechanics, also the plants and animals are treated in mechanist terms. Human physiology receives a similar treatment and in that sense Descartes sets out to propose different mechanical models for bodily functions like eyesight. Only the realm of the mind, reserved to man, the author of that feat of reason that is the new physics, is excluded. Indeed, his initial certainty can be established in these terms "I am thinking and whatever is thinking must exist, therefore I exist." From here he concludes that since any thing endowed with attributes is a substance and since one has the attribute of thought, one must be a substance. The source of these certitudes does not require an ulterior justification. Certitudes are given as "clear and distinct ideas" and their agreement with reality is guaranteed by God. Thus he constructs the world starting with two substances with distinct and incompatible attributes: the bodily substance and the mental substance.

Among the different attributes of each substance there is one that constitutes its essence, with the others referred to it. Following once more Galileo, he gives preeminence to a purely mathematical attribute, the extension, which is considered the essential attribute of the bodily substance. Other "bodily" attributes like the shape, size or movement refer to the extension. In other words, following to the letter the abstract models of classical mechanics identifies the world as a succession of instantaneous configurations of systems of material points that occupy successive positions in the mathematical space of Euclidean geometry. In spite of the fact that physics abandoned this simplistic vision of matter more than 100 years ago, it still has a strong influence and we could claim that it still forms part of the common sense idea of matter. In a good measure a central goal of this book is to contribute to update that concept. Concerning the mental substance, Descartes identifies thought as the essential attribute, with feelings, desires or imagination, referring to it. In order for a substance to be considered as such, it must be able to "exist by itself, that is without the need of any other substance". Once both substances are conceived as independent and defined in terms of incompatible attributes, the question of interaction of the substances arises, on which the possibility of any knowledge depends. Indeed if one describes the operation of the senses and the brain in purely mechanistic terms,as if they were clockworks, the connection between a state of the body and mind turns into a mystery.

With Descartes and the new physics of Galileo and Newton the new modern philosophy is born. Based on a still primitive comprehension of the physical world, its birth is marred by the unsolvable problem of dualism. All the philosophy after Descartes can be understood as a largely unsuccessful effort to resolve this problem. With this vision, with the sole exception of the human mind, everything is composed by inert matter whose essential attribute is extension. Nature, including the living organisms are mere mechanisms, are objects on which we can act with complete impunity for our own purposes. This vision continues to be the predominant one in our current forms of technological manipulation of the world. Nevertheless, to account for vital phenomena in the mechanist terms of Newton and Descartes has proved impossible. Today we know that even to explain chemical processes and the existence of atoms and molecules it is essential to use quantum mechanics which yields a much richer vision of matter.

Since the birth of mechanism a bifurcation takes place in the concept of nature: the one presented by science and the one yielded by our vital experience are divergent concepts. The same bifurcation takes place between philosophical approaches that either analyze nature in mechanist terms or ignore science because it cannot reveal the essence of being. Only a few philosophers of the 20th Century like Alfred North Whitehead or Hans Jonas undertook the task of developing a synthesis of scientific theory and lived experience.

Chapter 4

The downfall of the mechanist paradigm

4.1 Introduction

If one adopts a naturalist and physicalist viewpoint it is mandatory to understand how the notion of matter has evolved to analyze up to what extent the Universe can be hospitable to human concerns and needs of meaning. In the broadest sense, matter is anything that belongs in the field of study of physics. We have seen that the notion of matter of classical mechanics, based on the idea of corpuscles endowed with mass that interact with known forces is so limited that it only allows to build a universe of automatons that inexorably behave according to Newton's determinist laws. Such vision excludes any possibility of emergence of structures with causal capabilities or properties that do not reduce to those of their constituent parts. In particular we should abandon all hope of understanding the apparent efficiency of mental phenomena to produce physical phenomena. All our thoughts and intentions would be no more than an illusory appearance of purely mechanical developments that would be the ones that really explain why we act as we do.

In this chapter we will face the first great change in the notion of matter with the appearance of fields as material entities of continuous rather than corpuscular nature. With the notion of field a revolution in physics starts that continues to today and that we will discuss in detail in the following chapters. Parallel to the notion of fields the idea of Atomism develops from chemistry in the 19th Century. The dialectic process between the ideas of continuous fields in electromagnetism and discrete portions of matter in Atomism will have a superseding unification with the development of quantum mechanics as we will see in subsequent chapters. The efforts to explain both conflicting concepts in terms of Newtonian mechanics failed.

The world view presented by classical mechanics appeared to be, until the mid 19th century, totally satisfactory. Within its realm of applicability, mechanics does not encounter any inconsistency. If the theory was eventually superseded, it was due to the fact that it was incomplete. In the 19th century electrical and magnetic phenomena, known since ancient times, started to be studied systematically. At the beginning of the century scientists established that electricity and magnetism are related phenomena. In 1820 Hans Christian Oersted showed that a wire carrying a current generates a magnetic field. Towards 1830, Michael Faraday demonstrated that when a wire moves close to a magnet, electrical currents are produced. The final equations of the electromagnetic theory unified electricity and magnetism in a very elegant formalism which also included optical phenomena. They were developed by James Clerk Maxwell in 1873. In the following three decades many efforts were devoted to reduce electromagnetic phenomena to mechanical ones. It was thought that the former were due to the motion of a hypothetical substance called aether. The failure of these attempts demonstrated that mechanics was incomplete and there existed other forms of matter beyond the material particles, identifying the electromagnetic fields as the first material non-mechanical systems. In parallel with this, the understanding of the atomistic nature of matter gained strength, through the work of Dalton and developments in chemistry. The electromagnetic theory and atomism are at the origin of the two changes that determine the birth of modern physics and the supersession of mechanism: relativity and quantum mechanics.

4.2   The electric and magnetic field

Certain magnetic and electric phenomena have been known since time immemorial. It is known that Thales of Miletus made a series of observations about static electricity concerning substances like lodestone and amber, that rubbed with a piece of cloth could attract light objects like a feather, for example. Objects with permanent attractive or repulsive properties like magnetite were also known. There was little or no conceptual progress concerning electromagnetism till in 1600 the English physicist William Gilbert published *De magnete*, where he detailed careful experimental studies on electricity and magnetism. He even coined the word "electricus" which in New Latin means "like amber".

Charles-Augustine de Coulomb was the first scientist to formulate quantitative laws concerning electrostatics. In 1777 he invented the torsion balance to measure small forces, and used it to study the attraction and repulsion of charges. He noted that the force depended inversely on the square of the distance between two charges. He had discovered, about 100 years after Newton, that forces between electric charges behaved in a similar way as gravitational forces between masses. Two charges of the same sign repel and charges of opposite signs attract. Up to here all could be explained in mechanist terms concerning particles that interact through forces that obey known laws.

We can think of this phenomenon in an alternative way, that is shown graphically in the figure. A charged particle e_1 acts on its surrounding medium and creates in each point in space a vector field[1] called the electric field. The magnitude and direction of the electric vector at a given point are proportional to the force that a second charge e_2, assumed to be very small, would feel at that point. This second charge is called "test charge" as it is used to probe the electric field. The field produced by a charge at a given point is proportional to the magnitude of the charge and inversely proportional to the square of the distance between that point and the charge.

Magnetism was first identified as an independent phenomenon, associated with certain materials called magnets, that can exert attractive or repulsive forces among themselves or on certain other materials like iron or nickel.

As is well known, magnets have two poles and we can assign them opposite signs. Poles with equal sign repel and those with opposite ones attract. An interesting property is that if one cuts a magnet in half, one ends up with two magnets, each with two poles like the original

---

[1]A vector is a mathematical quantity that can be represented as an arrow and is characterized by a numerical value (the length of the arrow) and a direction in space (in which the arrow points). When one has one such arrow per point in space one talks about a *vector field*

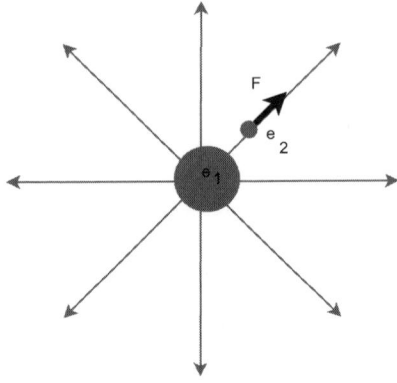

Figure 4.1: The field of a point charge.

one. One cannot have magnetic poles in isolation, they always occur in pairs. One can think that the magnet is composed by elementary magnets called dipoles.

The dipoles inside a magnet are ordered in such a way that all positive poles point towards the positive pole of the magnet and all negative ones towards the negative.

4.3   Electromagnetism

A great boost to the development of electromagnetism was provided by the discovery of electrical current. Towards the end of the 18th century Volta had constructed the first batteries, of some similarity to the ones we use today. They had the remarkable property of creating a potential difference that did not disappear when the battery leads were connected to circuits and they could maintain a current flowing for long periods of time.

Knowledge about magnetism was limited to magnets until, as we mentioned, in 1820 Hans Christian Oersted discovered that a wire carrying a current generates a magnetic field around itself and is capable of moving a compass needle placed in its surroundings. Many other experiments linking electricity and magnetism followed, due to André-Marie Ampère, Michael Faraday and many others.

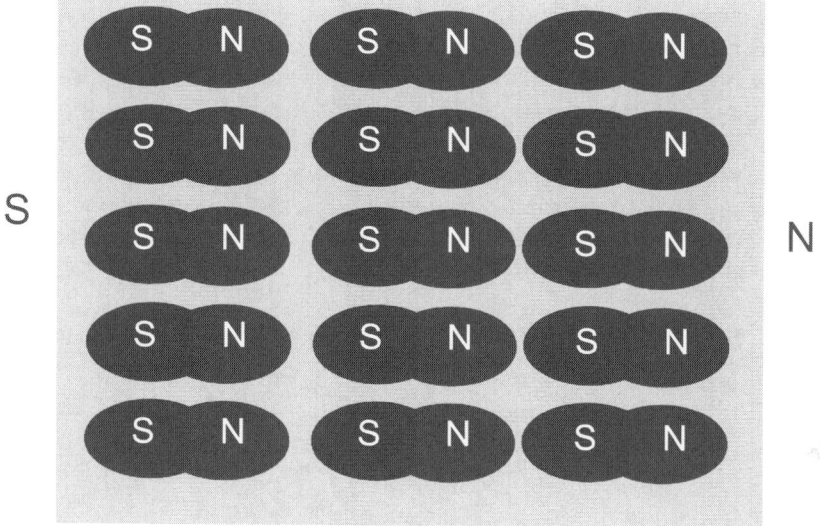

Figure 4.2: Elementary magnets (dipoles) form a magnet that behaves like having a magnetic North pole to the right and a South pole to the left.

Some of the forces generated by electromagnetic phenomena seemed to be velocity dependent and did not satisfy Newton's action and reaction law. This made the mechanist vision look less and less tenable. To these difficulties one has to add the ones stemming from optics and the strange physics of aether. The two problems would soon converge.

4.4 Electromagnetic waves

Let us assume that we move a charge that is the source of an electromagnetic field that was initially at rest. The motion of the charge will change the field surrounding it, producing a wave of changing fields that travels at the speed of light. The field close to the charge changes to follow the charge, but if we look far enough from it, the field still points in the direction stemming from the initial position of the charge. It takes time for the change to propagate. The laws of electromagnetism contained in

Maxwell's equations establish that electric fields that change with time produce magnetic fields and, reciprocally, time varying magnetic fields produce electric fields. In the example shown in figure 4.3, the charge that moves up and down, for instance like those in the antenna of a radio transmitter, produces variable electric and magnetic fields that propagate at the speed of light in the form of an electromagnetic wave. In the case of an antenna, a radio wave. Such waves carry energy and momentum and when they reach the antenna of a radio receiver they are able to move the electrons in it.

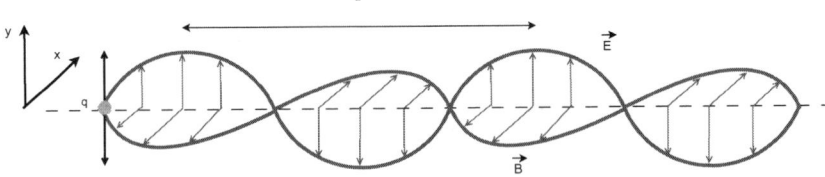

Figure 4.3: A charge q moving up and down produces an electromagnetic wave.

Even if the charges in the emitting antenna have stopped moving, the electrons in the receiving antenna will still be affected by the wave as long as the wave takes to go through to the location of the antenna.

The electromagnetic theory was cast in final form by Maxwell. However, as we mentioned, some of its pillars were laid by other remarkable scientists like Ampére, Faraday or Oersted. It not only unified electric and magnetic phenomena but optical phenomena as well. Indeed all phenomena of light propagation, like reflection, refraction, interference and diffraction are explained in terms of waves. The electromagnetic theory showed that radio waves and light waves propagate at the same speed and can be considered the same phenomenon, differing only in frequency, that is, how fast the electromagnetic field oscillates in the wave.

The electromagnetic waves largely behave like sound waves or waves traveling on a violin string, but unlike them, propagate in vacuum. In order to save the mechanist vision of physics, during several decades it was attempted to reduce electromagnetic theory to Newtonian mechanics assuming that the electromagnetic waves traveled on a material substrate, the aether. Electromagnetic waves were viewed as vibrations in the latter. In order to make sense of this analogy, the aether had to have contradictory properties. Several experiments determined that electromagnetic waves were transverse waves, the fields vibrated in directions perpendicular to the motion. Transverse waves develop in media that cannot be compressed but flexed transversally, like a rod of metal. Media that can be compressed, like gases, exhibit longitudinal waves in which vibrations happen along the direction of motion. This required the aether to be a solid. And given that the frequencies at which the fields change in light are very high, it had to be harder than any solid known. But it was a solid with the very odd property that did not interact with other solids, it was massless and it had no viscosity in order not to interfere, for instance, with planetary motion. Then came the question of motion. The Earth moves at a rapid speed with respect to the distant stars in its motion around the Sun and due to the motion of the latter with respect to the galactic center. Was the aether affixed to the stars or did it get "dragged along" with the Earth? If the aether was fixed to the stars and the earth moved through it, the light of stars coming from the direction into which the Earth is moving, would be faster than that of stars left behind. This could be easily ruled out experimentally, even in the 19th century. So the proposal was made that objects on Earth, like the lens of a telescope or a prism used to decompose white light into different colors, "dragged partially" the aether inside them. That partial dragging in turn accounted for why the speed of light in material media appeared smaller than in vacuum. However, in order to be consistent with experiments, it was required that the amount of dragging be different for lights of different frequencies, for which there was no natural motivation. Finally, more careful experiments like the famous Michelson-Morley experiment in 1887 ruled out the partial dragging theories entirely. With Einstein's theory of relativity all the attempts of explaining the undulatory phenomena in mechanical terms were abandoned.

4.5 The electromagnetism of Faraday and Maxwell

With the work of Michael Faraday and James Clerk Maxwell, electromagnetism reached its final form. In 1831 Faraday, inspired by the idea that if currents could produce fields, by reciprocity the latter should be able to induce currents, discovered magnetic induction: a variable magnetic field could induce a current in a wire. In 1845 he noted that magnetic fields could influence light when propagating in material media. The fact that light interacted with an electromagnetic field gave great support to the idea that light, electricity and magnetism were all related phenomena.

A simple experiment motivated in Faraday many of his ideas about the central role fields play. If one takes iron filings and sprinkles them around a magnet one notices that they align in curves that go from one pole to the other.

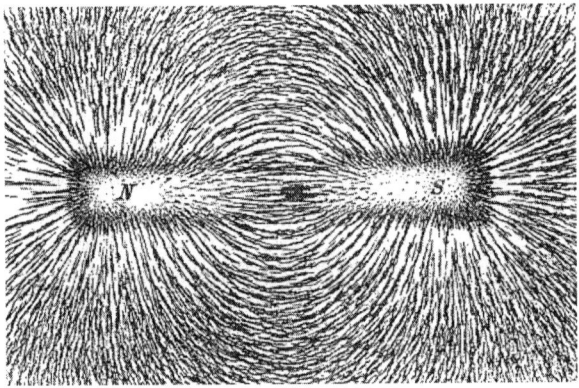

Figure 4.4: Iron filings align along the lines of magnetic field of a magnet.

The experiment convinced him progressively that there existed some property in space that generated electric and magnetic forces and perhaps gravity. His most important idea from a philosophic point of view was the acceptance of the fields as real entities. Towards 1850 he added arguments to the idea we already introduced when we discussed moving charges: that a field acts only locally and there does not exist an action at a distance. From his point of view, space was full of forces, being the particles the poles of maximum activity of these forces. He writes "Matter fills all space... or at least the space where gravity extends... because gravity is a property of matter, which depends on a certain force, and this force structures matter. According to this opinion matter is not simply mutually penetrable, but every atom extends, say, to the whole solar system, preserving however its own center of force." Faraday went even further, rejecting any idea that electric and magnetic phenomena were somehow associated with fluids without weight.

Maxwell gave its final form to electromagnetism. In particular, he established a set of differential equations that summarizes the whole set of electromagnetic phenomena. The equations describe the behavior of electric and magnetic fields interacting with charged particles. His equations were four. The first states that electric charges are the source of electric fields. The second states that there do not exist magnetic charges and therefore the magnetic field has no isolated sources. That is why magnets always have two poles (a pole in isolation would be an isolated magnetic source). The third equation establishes that variations in time of the magnetic field generate electric fields. The fourth and last establishes that both the motion of electric charges and the time variation of electric fields generate magnetic fields.

The above equations can be combined to show that electric and magnetic fields satisfy a wave equation and that the wave propagates without dispersion (it keeps its shape as it travels). Maxwell (Niven 2010) wrote "we have strong reasons to conclude that light itself (including radiant heat, and other radiations if any) is an electromagnetic disturbance in the form of waves propagating through the electromagnetic field according to electromagnetic laws."

Maxwell felt quite drawn by the mechanist prejudices and he claimed his equations were a "temporary instrument of research" and that they did not contain "even the shadow of a true physical theory". He continued trying to find a mechanist model of electromagnetism. His article "On physical lines of force" was an explicit attempt to present a mechanical model of a medium that connected electric and magnetic phenomena. But in his last works he finally abandoned the mechanical models.

The exit to all the problems generated by the mechanical models was nevertheless at hand. What was needed was to supersede the mechanical prejudices of trying to reduce every phenomenon to the motion of material particles, and to accept the physical existence of fields. The latter therefore acquire the status of physical systems on equal footing to the already known ones of systems of particles. This new vision relies, according to Laue (1913) "in a totally renewed understanding of the propagation of electromagnetic effects in vacuum; they are not supported by any medium nor do they give rise to an immediate action at a distance. The electromagnetic field in vacuum is something that has its own existence and an independent reality of any substance. Indeed, one must accustom oneself to this idea, but maybe this can be simplified if one remembers that the physical properties of this field, which are best given by Maxwell's equations, are more perfectly and exactly known than the properties of any substance."

In other words, the fields are not describable in terms of more fundamental systems of mechanical nature. In fact, Laue opposes the mechanist concept of substance normally associated with extended bodies, to the idea of field that nevertheless is treated as a "thing with its own existence". With this shift in point of view, physicists begin to supersede the traditional concept of substance, identified with a material mechanical substrate composed of particles with certain attributes. They go on to recognize something as real if it can act on other entities and to be modified by them. This change in the concept of material substance did not go unnoticed among physicists and philosophers of science. For instance more than a century ago Helmholtz said: "With respect to the properties of the objects of the exterior world, it is easy to see that all properties we can assign them, mean only the effects that they produce on our senses or on other natural objects... In all places we occupy ourselves with the mutual relationships between bodies... From this we conclude that in fact, the properties of objects of nature are not, in spite of their name, nothing "proper" of the objects themselves, in and of themselves, but they are always a relation to a second object (including the organs of our senses). The type of effect will depend, naturally, on the peculiarities of the body that produces is as well as those of the body on which the effect is produced." (Helmholtz 1867). But to renounce to absolute properties of bodies "in and of themselves" does not imply abandoning the objectiveness of knowledge that is not based on the absolute nature of things but in the events that they produce when they interact with a second object. Thus the fields are, given their very definition, defined by their effects, for instance on test charges, magnets or photographic plates and will produce events by accelerating charges or creating an image on a photographic plate. We encounter here for the first time the concept of a reality consisting of events that will eventually become a central notion of the new physical vision we present in this book.

Going back to the description of the evolution of physical ideas, one must consider fields as physical systems, as real as particles. More generally, we need to think of a charged particle as an object extended in space. The field that accompanies it in a given instant reflects basically the previous history of the charge. If one wishes to study the motion of a system of charged particles one needs to simultaneously describe the field that they create. In order to describe the state of a system in a given instant, and have all the information required to predict its evolution, it is not enough to give the position and velocities of the charged particles initially, as in classical mechanics. One has to add to that information the initial distribution of electromagnetic fields.

Maxwell's equations describe the structure of the electromagnetic field. They refer to the distributions of fields across all space and not just at the points in which there are particles or charges present, as one does in mechanics. They do not connect what happens here with what happens far away as mechanical forces do. The field here and now depends on the field in a neighborhood a few moments ago. Electromagnetic influences, in particular electromagnetic waves, propagate with a finite speed, equal to the speed of light. As soon as the implications of Maxwell's equations started to be understood, efforts to design a mechanist model became less interesting, and their results more and more disheartening. The last references to aether disappeared with Einstein's special relativity. This theory reaches a unifying vision of mechanical and electromagnetic phenomena using electromagnetism as a starting point and modifying Newton's mechanics.

4.6   Electromagnetic properties of light

Maxwell discovered the wave nature of electromagnetic fields through his equations. He was able to compute from fundamental constants, determined independently studying electric and magnetic phenomena, the speed of propagation of the waves, and noted that it coincided with the speed of light. This led him to conclude that light had an electromagnetic nature. As we mentioned before electromagnetic waves are transverse, meaning that the electric and magnetic field point in directions transverse to that of propagation and their directions differ by 90 degrees. The direction in which they point can remain constant in time. In that case one says that the electromagnetic wave is *linearly polarized* (as in figure 4.3). Or their direction can rotate, in which case one says that the wave is *circularly polarized* (as in figure 4.6).

A characteristic property of a wave is the frequency. In the case of light the frequency is the inverse of the time that the electromagnetic field at a given point takes to return to a given value. The higher the frequency, the faster the field is oscillating. The frequency of light is associated with the color. If one could take a snapshot of the electromagnetic field at a given time as a function of space, one gets a sinusoid. The distance between two successive crests (or valleys) in such a sinusoid is called *wavelength* and we denote it with λ. The higher the frequency the smaller the wavelength and vice-versa. The maximum absolute value of the field is called the *amplitude* of the wave.

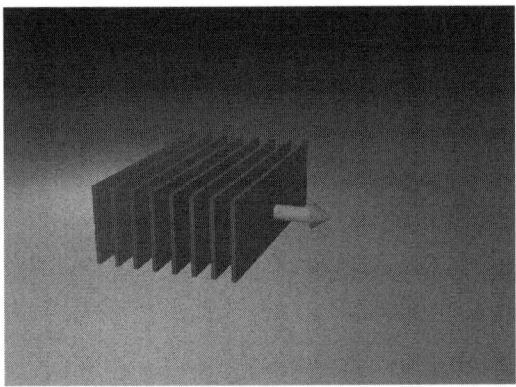

Figure 4.5: A representation of a system of plane electromagnetic waves. The propagation direction is indicated by the arrow and the planes represent the regions where the electric field is maximum.

The simplest type of electromagnetic waves are the *planar waves*. In them, all points in a plane perpendicular to the direction of propagation have the same value of the fields. Plane electromagnetic waves do not have to be polarized in a plane, they can be circularly polarized as well, as the figure shows. In a circularly polarized wave the electric field vectors have a constant magnitude but their direction changes as a function of time in a rotational manner. In a planar circularly polarized wave the electric field, from plane to plane, has a constant strength while its direction steadily rotates. Since the electric and magnetic field directions are separated by 90 degrees at all time, if one were to draw the magnetic field it would also describe a helix.

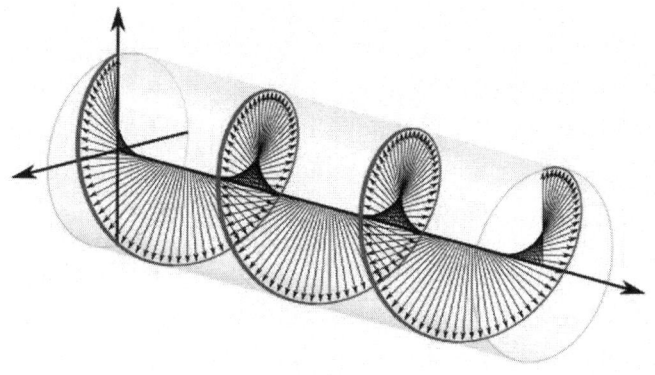

Figure 4.6: A circularly polarized wave.

4.7 Interference

If, at a given point, a point charge q_1 creates a field $\vec{E}_1$ and another point charge q_2 generates a field $\vec{E}_2$, the combined field of both charges at that point is $\vec{E}_1 + \vec{E}_2$. Due to this property one says that one can superpose electric (and magnetic) fields. Notice that since the fields are vectors one has to add them appropriately, taking into account not only their magnitude but the directions as well.

Interference occurs, for instance, when one superposes two waves with equal frequency and amplitude. Let us assume we have two electromagnetic waves in a region of space traveling in the same direction. If both waves at a given point have the maximum value of the electric field A, then the field at that point will be $2A$ and we will say that the waves interfere *constructively*. However, if at a given point one wave has the maximum value A and the other has the minimum value $-A$, then when one adds the fields one gets zero. The waves have interfered *destructively*.

The two slit (or two pinhole) experiment of Young is particularly interesting since we will see it plays an important role in quantum me-

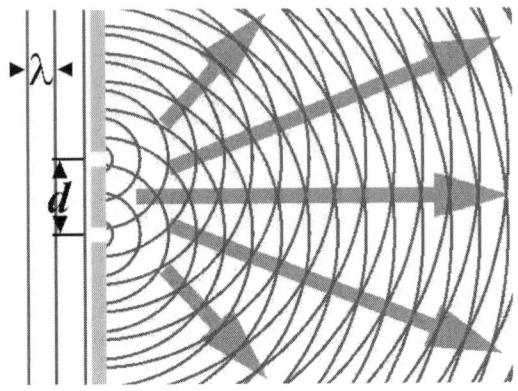

Figure 4.7: The double slit experiment of Young.

chanics later. Let us suppose that a plane wave hits a barrier with two slits. As the figure shows, every point of the wave that makes it through the slits behaves like the source of a new wave that propagates away from the slit in all possible directions with the same speed and frequency as the incident wave. One ends up with two waves that are not planar anymore, but are cylindrical, centered in each slit. In a cylindrical wave the fields take the same values along cylinders. In particular the "wavefronts" (the points where the field is maximum) lie on cylinders centered at each slit for each wave. The superposition of the two cylindrical waves creates a complex pattern in space of points where the waves interfere totally or partially destructively and constructively. If one were to take a screen that is parallel to the barrier in the region where the waves are interfering, the light would produce a pattern as shown in figure 4.8. One gets a series of bars of light corresponding to the region surrounding the place where the waves interfere fully constructively alternating with bars of darkness which are centered in the places where complete destructive interference takes place.

Young's experiment reinforced the belief in the wave nature of light. There was no natural way for Newton's corpuscles to traverse the slits and produce the type of pattern Young saw. Notice that the interference pattern only happens if light goes through both slits. If one covers the second slit the intensity of light in a given point of the screen is I_1.

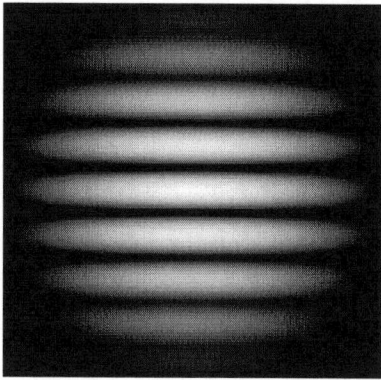

Figure 4.8: A picture of the pattern of a double slit (more precisely double pinhole, therefore the finite horizontal extent of the lines) experiment. Image: T. Weitkamp.

If one covers the first slit the intensity at the same point is I_2. The intensity with both slits open is not the sum $I_1 + I_2$, which is what one would expect from Newton's corpuscular theory. The prediction of electromagnetism is different. The intensity at a given point depends on the square of the electric field at that point. An in this case the field at each point is given by a superposition of the fields stemming from both slits. So the intensity due to both slits is not simply given by the sum of intensities, due to the square there exist additional terms. One can therefore have total destructive interference at the point in which case the field and the intensity is zero. Or total constructive interference where the intensity is maximum. Or values in between. This leads to the smooth pattern of stripes seen in the figure.

4.8 Atomism

Within mechanism dwelled two different visions of matter: the atomistic one and the continuistic one. The atomistic one had a long history going back to the Greeks five centuries B.C. Without experimental confirmation, preference for which vision was divided among thinkers in the 17th and 18th century . Although conceptually mechanics made refer-

ence to particles that were treated as points, such treatment could be
considered an idealization of extended configurations like those of fluids
and solids in a simplified way. Descartes was in favor of the continuist
vision whereas Newton leaned towards the atomistic view of the Greeks.
In spite of Newton's influence and the general acceptance that some sort
of corpuscular structure may underlie matter, the corpuscular point of
view did not yield scientific insights during the 18th century. Neither
experiments nor theoretical developments helped advance its case. In
fact, the success of Euler formulating hydrodynamics of continuous fluids in terms of differential equations appeared to render the problem of
the corpuscular underlying nature of matter into purely metaphysical
terms.

The principles of atomism reappear with Lomonosov, who anticipated the kinetic theory of gases and with John Dalton in the context
of chemistry. He established that all bodies of observable size are constituted by a large number of extremely small particles, called atoms,
linked by attractive forces. He added that the ultimate particles that
compose a substance like water must all be identical: "every particle of
water is like every other particle of water; every particle of hydrogen is
like every other particle, etcetera" (Dalton *et al.* 1893/2011). Dalton's
hypothesis originates mainly in the study of chemical reactions and in
the observation that the elements combined according to their weight
in definite proportions, give rise to substances.

In spite of emanating from chemistry at the beginning of the 19th
century, these ideas are not impactful till the second half of the century,
when they are adopted to explain crystal structures and microscopic
properties of electric and magnetic materials. But the atomistic ideas
resulted most fruitful in the development of the kinetic theory of gases,
which allowed to progressively establish the microscopic basis of thermodynamics, in particular concepts like temperature and heat.

A contribution from atomism that had profound consequences in the
changes in physics that would happen in the 20th century, particularly in
quantum mechanics, was the introduction by British natural philosopher
Richard Lamming of the hypothesis of an elementary charge, later to
be called electron, confirmed experimentally by Thompson in the last
years of the 19th century. Atomism, which had started as a philosophical
doctrine, became now part of experimental science and would lead to
one of the more important avenues of research of 20th century physics,
the search for elementary particles that compose the atoms of all known

elements.

The atomic view of matter seemed to match well with the ideas of classical physics. After all, Newtonian physics deals naturally with point particles. It would later be verified that this belief is actually false and in order to describe the behavior of atoms and the elementary particles that compose them it will be necessary to introduce the most revolutionary of physical theories known up to present: quantum mechanics.

4.9 Conclusions

Summarizing, we have seen that electromagnetic phenomena are described in terms of fields and charged particles. Fields manifest themselves through their effects on charged particles or magnetic dipoles, like iron filings, and they extend throughout space. Electromagnetic waves, like the radio waves, convey energy and information between distant points and propagate without any need of a mechanical substrate be it either a fluid or a system of particles. Light is a particular case of electromagnetic wave.

Fields are, as Laue put it, a "thing with its own existence". Something is recognized as material if it can act on other entities and is modified by them. A new notion of object arises in electromagnetism, that is recognized only by its effects on other objects. As Helmholtz (1867) put it: "mean only the effects that they produce on our senses on other natural objects... In all places we occupy ourselves with the mutual relationships between bodies... From this we conclude that in fact, the properties of objects of nature are not, in spite of their name, nothing 'proper' of the objects themselves, in and of themselves, but they are always a relation to a second object (including the organs of our senses)."

We encounter here for the first time the concept of a reality consisting of events that result from those interaction processes referred to by Helmholtz that will eventually become a central notion of the new physical vision we present in this book.

Concerning Atomism, the initial mechanist idea that the properties of macroscopic matter result from the combinations and motions of small atomic particles soon proved insufficient. Atoms had to be systems composed by charged particles with the same properties for all atoms of the same element that could hardly be explained in mechanical terms. Only with the advent of quantum mechanics a satisfactory understanding of atomic structure and how atoms combine to give rise to more complex substances arose.

Chapter 5

The scenario becomes an actor: space-time as a form of matter

We will see in this chapter that with the birth of the theory of relativity developed by Einstein between 1905 and 1916 two closely related conceptual changes took place: 1) The very space and time that were part of the scenario on which particles and fields evolved becomes a material element. Instead of an abstract space described by Euclidean geometry and a universal time that flows in the same way for everyone, a notion of space-time emerges that can propagate energy and information. Space-time does not have a fixed geometry anymore but its geometry is affected by other forms of matter. 2) The role of events, first noted in the discovery of the notion of fields, becomes more and more important, being one of the goals of relativity to specify the possibility of one event influencing another. As we already noticed we never observe a field directly but through its manifestations, like a field influencing the motion of charges, or light affecting the retina of our eyes. Our senses establish a causal relation between events that take place in our receptor organs, the eye or the ear and the sensations they trigger in our mind that very likely are also events in our brain. We would therefore have "direct" access to certain events related to our perceptions. With the introduction of the notion of event and that of relativity, the role of observations becomes more important. Indeed, the concept of objectivity changes with relativity. It does not consist in having the same properties for all observers, but in the possibility of relating observations of different agents according to well defined and precise rules.

Objectivity resides in the possibility of relating and coordinating the set of all perspectives that an object offers to different observers, that is all the events that the object produce in the sensory organs or in measurement instruments. Finally we note that space-time, just like fields, acquires reality through its manifestations. The manifestation of space-time will be, as we shall see, the way in which events causally interact among themselves. Space-time defines which events can have effects on others.

5.1   Special relativity

Let us recall that for Galileo and Newton time intervals and distances between points are absolute concepts, that is, they do not depend on the state of motion of whomever is measuring them. The Newtonian universe is composed of particles whose positions and velocities vary with time obeying pre-established force laws.

As we all know velocities depend on the observer. This property is a direct consequence of the invariance of spatial and temporal intervals in classical mechanics. If two cars advance along a road moving in the same direction with velocities $V_1 = 90km/h$ and $V_2 = 120km/h$, the second gets away from the first with a velocity $V_{12} = V_2 - V_1 = 30km/h$.

Let us address how Einstein approached the issue of relativity. His first observation is that *fields are as fundamental as mechanical systems*, as we have argued in the previous chapter. As a consequence, both the laws of mechanics and the laws of electromagnetism should have a similar status in physics. His second observation is that *the relativity principle, first introduced by Galileo for mechanics, should be valid for all fundamental laws and not only those of mechanics.*

To understand the relativity principle one needs to define frames of reference. A frame of reference attached to a body serves to describe the position of points relative to the body. It is also known as a coordinate systems, and is composed by axes (lines) emanating from a point known as the origin.

An inertial frame of reference is a frame of reference such that particles free of forces move uniformly in a straight line, simplifying inertial frames are systems that are not accelerated by any external force and are not rotating with respect to very distant astronomical objects. There does not exist any way to distinguish if an inertial system is at rest or in straight uniform motion. Einstein's relativity principle establishes that the laws of physics are the same in all inertial systems of coordinates.

Einstein's third observation is as follows: *if the relativity principle we just discussed is valid for Maxwell's laws of electromagnetism, then the speed of propagation, which is a prediction of such laws, should be the same in all inertial reference frames.* This observation is verified by experiment: the speed of light is always the same, no matter what the speed of the emitter and/or the receiver is.

Einstein noticed that from these observations a first conclusion immediately follows: the notions of space and time cannot be the ones of classical mechanics. In fact, as we mentioned above, from these notions follow that the velocity of an object respect to certain observer depends on the speed of the observer and they would lead immediately to a contradiction with the third observation: if for an observer the speed of light is c in another system that is moving with respect to the first it would take a different value, for instance $c + v$, and not c. If one adopts as fundamental the axioms of classical mechanics and assumes the invariance of space and time intervals then Maxwell equations cannot be valid in any inertial reference frame and the speed of light would depend on the frame. In order to have a description of the electromagnetic phenomena valid for any inertial observer one needs to introduce new notions of space and time.

Even though at the time of Einstein there were experiments that suggested that the speed of light was independent of the reference frame used (like the celebrated Michelson–Morley experiment (1887)), the choice in favor of keeping Maxwell's equations invariant was made by Einstein purely due to theoretical reasons.

Galileo's relativity is based in principles apparently indisputable from the point of view of current common sense, about the measurement of distances, lengths and times. Just like Galileo did in his time, Einstein charges against common sense and adopts a solution that is apparently unacceptable and that revolutionizes our concepts of space and time. Again, as in Galileo's argument of the tower providing support for the notion of inertia in classical mechanics, the creation of a new theory goes hand in hand with a job of interpretation. Its objective

is to show that the new axiomatic systems are consistent and account
for known experiments in spite of the apparent contradiction with the
predominant common sense at the time. Implicitly, Einstein changed
the definition of an inertial reference frame. Now a system is inertial
if a) particles free of forces move uniformly in a straight line and b)
Maxwell's equations are valid.

5.1.1   Relativity of simultaneity and lapses of time

Let us explore the consequences of the principles we just stated. A first
simple conclusion is that simultaneous events in a system of reference are
not necessarily so in another. In fact, let us suppose we wish to compare
the observations made in a ship, moving at a speed v with respect to
the pier, with those made from the pier. Suppose light is emitted from a
light bulb in the ship, situated at equal distances from the bow and the
stern. For someone traveling in the ship, the light front will reach the
bow and the stern at the same time given that light propagates at the
same speed in all directions and bow and stern are at the same distance
from the light bulb. From the point of view of the pier, the speed of
light is the same as that on the ship, it is not influenced by the fact
that observed from the pier, the light bulb is moving. However, from
the point of view of the pier, the bow moves away from the light and
the stern towards it. Therefore the light will have to travel a longer
distance to reach the bow than to reach the stern. Therefore it is not
true that light will reach both at the same time, it will reach the stern
earlier than the bow.

This result, so easy to understand, yet so surprising, contains the
essential elements of Einstein's relativity theory. The same events: the
arrival of light at the bow and stern are simultaneous for one observer
and not for another. This implies the demise of the famous Newtonian
absolute clock for which time evolved unaltered no matter where in the
universe we were. In the example, one of the events is simultaneous with
the other for one observer whereas for the other one event occurs previously to the other. Therefore past, present and future do not have an
absolute sense any more. If two friends that walk past each other in the
street ask themselves if at this precise instant a supernova is exploding
in a distant galaxy, the answer could be yes for one of them and no for
the other. The small difference in velocity for the two friends walking
towards each other puts them in different reference frames. Therefore

an event simultaneous in one of them will not be in the other, it could happen weeks later. There does not exist an "instantaneous universe" constituted of "everything that exists now". The mechanist conception of a universe that evolves from instant to instant becomes unsustainable.

With respect to this idea Schrödinger (1944/2012) said "I suppose it is this, that it meant the dethronement of time as a rigid tyrant imposed on us from outside, a liberation from the unbreakable rule of before and after. For indeed time is our most severe master by ostensibly restricting the existence of each of us to narrow limits —seventy or eighty years, as the Pentateuch has it. To be allowed to play about with such a masters programme, believed unassailable until then, to play about with it albeit in a small way, seems to be a great relief, it seems to encourage the thought that the whole timetable is probably not quite as serious as it appears at first sight. And this thought is a religious thought, nay I should call it the religious thought"

If there is no notion of absolute simultaneity and different observers assign different times to the same events, we need to review what we mean by "time". For a physicist, "time" is what clocks measure. This leads us to ask: what is a clock? Simplifying things a bit, we can think of a clock as a mechanism where a phenomenon repeats an indefinite number of times. The old pendulum clocks, the modern atomic clocks based on atoms oscillating or even the human heart are good examples. In fact, Galileo studied the uniformity of the period of the pendulum (independently of the amplitude of oscillation) using his own pulse as a clock.

Einstein had toyed with the hypotheses that serve as starting point for relativity for years. Then in 1905 he observes that they are only compatible if one revises the notion of time. Clocks behave differently when they are in motion that when they are at rest. In order to study how clocks are affected by motion, one can use a "light clock". This is a very simple type of idealized clock (which nevertheless is very difficult to realize in practice). It consists of two parallel mirrors mounted a distance apart and a pulse of light bouncing back and forth between them, as the figure schematically shows. Time is measured counting the number of bounces ("ticks") that the light suffers during the interval of interest. Let us assume we have one of these light clocks at rest in our lab, whereas another is in motion. We would like to know if in both clocks, light will bounce at the same time. From our perspective, in the clock at rest, every time the photon traverses the separation of the two

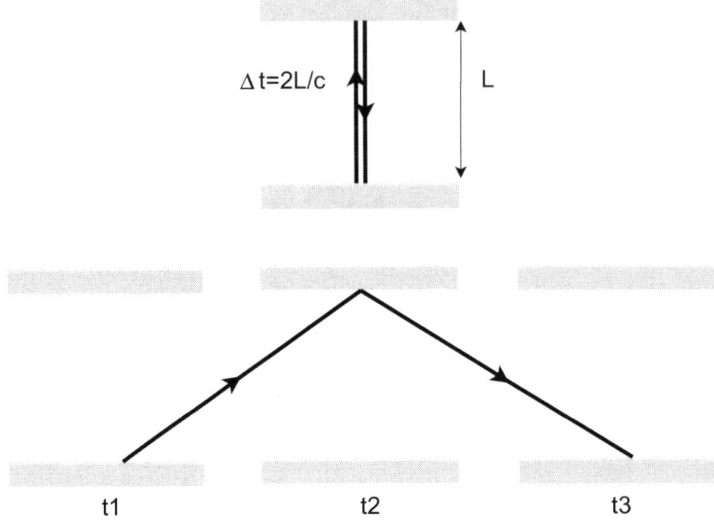

Figure 5.1: The idealized clock discussed in the text. And a depiction of three instants of time and how light propagates when the clock moves, in this case sideways.

mirrors we will hear a "tick".

However, in the moving clock, although the light travels at the same speed, the mirror will be receding away from the photon, leading it to travel a longer distance before ticking. As the figure depicts, the photon in space-time traverses a diagonal path of longer distance than the separation of the two mirrors at rest. Therefore the moving clock moves slower from our perspective. Notice that the slowness of the clock increases the faster the clock moves. For the types of speeds involved in trains or even planes, the slowing of the clocks is negligible (although the slowing of a clock in a plane is measurable using high precision atomic clocks, this is known as the Hafele and Keating (1972) experiment).

One could ask if this is some fictitious effect due to the peculiar clock we are considering. The answer is negative. Let us place our clock in a train in uniform motion, together with an ordinary wrist watch and an atomic clock. Will those two read different times? The principle of relativity says that we cannot distinguish if the train is in uniform motion. If the clocks read different times, we would have a way to tell the train is in motion. Therefore all clocks must behave in the same way.

As we have stated, relativity allows to relate the different perspectives offered by a given set of events. In the case of the "light clocks" the bounces of the photon on the mirrors. When we change the reference frame, the perspective changes. For instance, if we study the clocks instead of from the lab frame from the train frame we will see that now it is the clock in the lab the one that runs slower. Therefore from one perspective the observer in the lab ages faster and according to another it is the observer in the train who does. Could not this lead to contradictions? As the reader may anticipate the answer is negative. An attempt to create a contradiction is the well known "twin paradox". Twin siblings Alice and Bob decide to prove relativity wrong. They board spaceships and synchronize their stopwatches in an instant in which the spaceships, moving at enormous speeds, cross by each other. As they speed away from each other both think that the other is aging more slowly. To create a paradoxical situation they need to meet each other again to verify who aged faster. For this, Alice fires her ships thrust engines to reverse her motion and go to meet Bob. When they meet they discover that Alice is younger. The symmetry of the situation was lost when Alice decided to fire her thrust rockets. In that process she underwent acceleration, so we cannot assume she was at rest (or in uniform motion) all the time. That is, there is no contradiction to the relativity principle since the latter only applies to inertial reference frames and not to systems that accelerate relative to each other. This experiment only confirms that time flow depends objectively on motion. To explain the experiment entirely one needs to consider general relativity, where arbitrary reference frames are allowed.

The skeptical reader may claim: OK, this is what relativity predicts and it is self-consistent, but is it what it is observed experimentally? Indeed these types of effects are routinely observed by physicists in systems that move at speeds close to the speed of light. An example is given by the disintegration of elementary particles. There exist many unstable particles that moving at slow speeds disintegrate in fractions of second. Particle accelerators can get these types of particles to 99.9% of the speed of light and their lifetimes increase by up to factors of ten. That is, while moving, these particles take much longer to disintegrate. A similar effect is observed in some types of cosmic rays that reach the

Earth from deep space, the only reason they can make it to Earth is because their lifetimes are extended by moving at very high speeds.

One can show that spatial intervals are not invariant either. An easy way of checking this is observing that if time intervals depend on the observer and spatial intervals do not, then light would not have the same speed for all observers.

5.1.2   Neither space nor time: space-time

Relativity concerns events organized and coexisting in a space-time. Neither the spatial interval between two events nor the time lapses between them are invariant, as used to be the case in classical mechanics. Both depend on motion. Although the division between space and time is not absolute anymore, in special relativity space-time has a fixed structure: its geometry. The geometer Hermann Minkowski was the first to realize that space and time must be considered as a single entity, a four dimensional space-time, that requires the introduction of a new idea of "distance" between events in space-time. The Minkowskian distance, also known as relativistic interval, is independent of the reference frame in which it is computed. Just like in three dimensional space the distance between two points does not depend on the set of Cartesian axes used to compute it.

The square of the relativistic interval between two events is defined by minus the square of the speed of light times the square of the time interval plus the square of the spatial distance between the events. Given an event E_1 that occurs here and now, one can classify any event E_2 by computing its interval with E_1. One says that an event is to the future of E_1 if the interval is negative and t_{12} is positive. This in particular means that one can send signals from E_1 to E_2 and therefore E_1 could influence E_2. In the same line we say that E_2 would be in the past of E_1 if the interval is negative and t_{12} is also negative. That means that in this case it is E_2 who could send a message to E_1. Notice that although t_{12} is a frame dependent quantity, its sign is not, so the statements about the future and past are invariant.

The importance of events to the future and past of E_1 can be highlighted by contrasting them with events E_2 for which the interval s_{12} is positive. In the former case one says that events are "time-separated" whereas in this case one says the events are spatially separated. Events of that sort cannot send signals between them. A simple way of picture

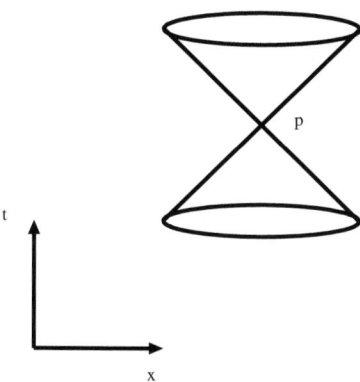

Figure 5.2: Light cones separate the points of space-time with respect to a given point P into three regions, one time-like to the future and past of it, one space-like to the sides and a cone of points whose space-time distance to P vanishes.

them is that they are too far apart for anything to communicate between them within their time separation. One would need to propagate faster than light for communication between space-like separated events to occur, and as we saw, this is not possible.

At this point it is a good idea to introduce the concept of light cone which is the surface in space-time consisting in all points whose interval with a given point is null. These are the points that are swept in time by a light wave emanating from the point in question, which is at the vertex of the cone (it is really a higher dimensional hyper-cone, its sections are given by spheres rather than circles). The invariant space-time interval, as we have seen, leads to the impossibility that certain kinds of events could influence others. If an interaction happens or not is something absolute, and this is a basic observation for the analysis of all elements that constitute reality. Events can therefore be ordered according to the possibility they have of interacting with each other, constituting what is known as the causal structure of space-time. Although neither space nor time have an absolute character anymore, the causal structure is absolute. Space-time therefore describes a network of events and their causal relations.

Space time has a geometry with parallels to the usual Euclidean one we are accustomed to in space, but the "distances" are measured with the invariant interval we discussed. Given two points that are spatially separated, the curve that minimizes the invariant interval between them is a straight line. So in Minkowski's space-time we still have the notion of a straight line and of uniform straight motion. In Minkowski's space-time we have an absolute distinction between uniform and accelerated motions.

5.2   General relativity

5.2.1   Origins

One of the most remarkable contributions of Newtonian physics was to provide a simple law to describe gravity. The latter is the most universal of the forces of nature; all physical objects without exception produce gravitational effects on other objects and in turn feel the gravitational attraction of them. The universal law of gravitation of Newton allows to compute the force with which two massive particles attract each other. The force is larger the larger the masses in question and diminishes the farther the particles are apart. More quantitatively the law establishes that the gravitational attraction is proportional to the product of the masses of the particles and inversely proportional to the square of their distance R,

$$F = G\frac{mM}{R^2} \tag{5.1}$$

where G is Newton's universal constant and m and M are the values of the mass of the two particles. This is the law that Newton himself used to study the motion of planets and comets and even today is used to compute the flight paths of spaceships used in space exploration, as well as the calculation of ballistic projectiles by the military, or even the simple motion of a soccer ball.

Einstein observed that this law would contradict relativistic principles. Let us recall that when an electric charge moves, it influences first the electromagnetic field nearby and its influence on distant charges only happens when the electromagnetic wave generated by the first charge reaches the second. In contrast, gravity as described by Newton propagates instantaneously. For instance, if the Sun were to explode today, it would take eight minutes for us to see it, as that is the time light takes to travel from Sun to Earth. But gravitationally we would feel the effect instantaneously. This instantaneous propagation of gravity contradicts relativity, since as we argued, it does not allow the possibility of emitting signals that travel faster than light.

Einstein was aware of this problem and the consequent need to change Newton's law of universal attraction. The latter would eventually be superseded by the general theory of relativity, but the main driving force for the development of that theory was the desire to get rid of any preferred system of reference, as the inertial systems.

Both in Newtonian mechanics and in special relativity, the inertial reference frames play a central role. They are privileged systems with respect to which the physical laws take their simplest form. This special role of inertial systems depends ultimately on the fact that although the uniform motion of a system cannot be detected by experiments performed on the systems in question, acceleration is absolute. Nothing prevents from describing physics using non-inertial frames, for instance inside a car that is taking a turn. But new forces appear,in addition to the usual forces, called fictitious forces. In the case of the turning car the driver will feel a sideways pull away from the turn known as centrifugal force in the frame of reference of the turning car.

The mass of a particle can be viewed as a measure of its resistance to being accelerated. But accelerated with respect to what? The answer is simple: with respect to inertial frames. In spite of the simplicity, the answer is not entirely satisfactory, and even Newton was aware of this. Without a better answer, Newton appears to have attributed to absolute space the ability to resist acceleration. But why would space be incapable of distinguishing between systems with different velocities when it can distinguish different accelerations? In what sense can space, which is a pure mathematical concept, a receptacle where particles reside, have the ability to act generating forces in non-inertial frames? Einstein never accepted the idea of an absolute space. He argued that: "It conflicts with one's scientific understanding to conceive of a thing which acts but cannot be acted upon." That is the case of the absolute geometric space of Newtonian physics. In spite of his rejection of absolute space, Einstein could not get rid of it with his special theory of relativity. One still has to refer to privileged inertial reference frames with respect to which we measure acceleration. One keeps the idea of space-time as a mere vessel for the events taking place in it, remaining

unaltered by anything external, and described by a geometry essentially
Euclidean with the distance between events given by the Minkowski
invariant interval.

Einstein observed that gravitation should be related with inertia.
The reason is very simple: phenomena seen from an accelerating system
are equivalent to those performed in a inertial frame with a gravitational
field present. When the flight of the Space Shuttle starts, astronauts feel
their weight increases by a factor of five. We know that such increase
is due to the acceleration of the Shuttle at takeoff. Conversely, bodies
in the presence of a gravitational forces fall with the same acceleration
and one can eliminate the acceleration by going to a reference frame in
free fall. That is precisely what happens when a spaceship is in outer
space. Inside the cabin astronauts appear to be suspended in space because they all, including the ship, move in the same orbit with the same
acceleration. So going to an accelerating reference frame it is possible
to create or eliminate gravitational effects. That means that including
accelerating frames is connected to including gravity. That is the basis
of the equivalence of all reference frames: any observer irrespective of
its motion can claim that it is at rest if it includes gravitational effects.
Through the inclusion of gravity, the general theory of relativity allows
to treat all systems of reference democratically, in totally equal footing.

Free falling systems, even in the presence of strong gravitational
fields, are totally equivalent to inertial reference systems. Each of those
free falling systems constitute local inertial frames. Given a chosen
event, one can always imagine a free falling frame locally that behaves
like an inertial frame. Any other system in uniform motion with respect
to that inertial frame is also inertial. Viewed from this perspective,
special relativity has expanded its range of action to include intense
gravitational fields. However, its range of action has become local. Let
us explain why we are insisting on the local aspect. The gravitational
field of a body like the Earth is not strictly uniform. The forces of
attraction point towards the center of the Earth. In the free falling
system, for instance, an elevator cab whose cable broke and is falling to
the ground floor of the building, or the Apollo spacecraft cabin before
it opens its parachutes and enters the atmosphere, gravitational effects
are almost completely eliminated. If one takes two particles initially at
rest, for instance, one against each wall of the elevator, they will remain
almost at rest. However, as time passes an observer inside the elevator
will notice the particles moving toward each other. This is because the

elevator walls are falling parallel to each other whereas the particles are falling radially towards the center of the Earth. Given the long distance to the center of the Earth compared to the size of the elevator cab, the trajectories will be approximately parallel. But not exactly. These types of effect can be made arbitrarily small by limiting the size of the cab and the length of time of the experiment. This is what is meant by "locally" in space-time.

As gravity is a universal force that couples with all forms of energy in the same way it is possible to eliminate the gravitational effects going to a system of reference that falls with all the particles. This observation about the universality of gravity led Einstein to suppose that particles under the action of gravity behave as free particles moving on a curved space-time. In other words, as all particles no matter what is their mass follow the same trajectory, Einstein was led to conclude that any particle when it is only subjected to a gravitational force follows a trajectory that is the closest thing there is to a straight line in a curved space time.

General relativity is a theory where space-time is generically curved and in which the physics is invariant under any changes of reference frame. It will also be a theory of gravity, where the latter will not be represented by a force, but by the curvature of space-time. As such it is unique among the fundamental interactions, since all others are described by a force living in a background space-time. As we discussed in the introduction, we will not attempt to cover in detail general relativity, just to introduce some basic elements.

To deal with general relativity we therefore need to learn a bit of how to handle curved geometries. Our information about the curvature of space-time only comes from our experiences, all carried out *from within* that space-time. We do not know anything about a hypothetical space of more dimensions that would include our space-time. If such a thing even existed, we could not access it. An analogy is our situation in the Earth. An astronaut flying above the Earth immediately realizes it is curved. However, it is much harder to establish such a fact if one is on the ground. To determine that a space is curved from measurements made inside the space is to determine its *intrinsic* curvature, without reference to any external space from which it could be seen. Intrinsic properties only depend on measurements made within the space. Identical intrinsic spaces can be seen in different ways when embedded in spaces of higher dimensions. For instance, a two dimensional sheet of space can be embedded into a three dimensional space as a plane, or

rolled up in a cylinder. Intrinsically nothing changes. For instance, if
one draws a triangle on that piece of paper, its dimensions, angles, etc.
do not change if one rolls up the paper into a cylinder. The plane and
rolled up pieces of paper cannot be distinguished from each other from
measurements made within the piece of paper[1]. On the other hand, if
instead of paper we were dealing with rubber, in addition to rolling it up
into a cylinder we could wrap it around a sphere. In that case, however,
the material is stretched and distances and angles would change. Such
changes are intrinsic, they could be determined by an insect living on
the sheet of rubber by measurements done on the piece of rubber itself.

To have a notion of intrinsic curvature one needs to at least be able
to measure distances in the space, and to do so in arbitrary coordinates.
In regions where gravitational fields are weak, for instance portions of
space far away from gravitating bodies, we can use the notion of space-
time interval of special relativity in suitable coordinates.

A geometric notion that has intrinsic meaning once one has a notion
of distance is the idea of *geodesic*. Given a pair of points in a space, a
geodesic is a line that connects them that has the shortest length. For
instance, in a sphere, the geodesics are the maximum circles. In a plane,
geodesics are simply straight lines. In curved spaces there is no notion of
a straight line and geodesics are the closest thing one has to them. The
notion only requires measuring distances, which is a concept intrinsic
to the space one is considering. The curvature[2] is another notion that
can be determined intrinsically. If we are two dimensional beings living
on a surface, we can consider all geodesics stemming from a given point
and construct a "circle" by measuring out a fixed distance along each
geodesic. If we measure the length of the circle and its radius, we would
get $2\pi r$ if we are in a plane, something smaller if we are on a sphere,
or something bigger if we are on a surface that resembles a saddle.
These ideas can be readily generalized to higher dimensional spaces.
The notion of geodesic does not depend on the number of dimensions.
Similar operations to the ones we carried out with the circle can be done

---

[1] An apparently obvious way to distinguish the two situations would be to attempt
to circumnavigate the piece of paper. That would correspond to a highly non-local
measurement. The laws of physics in most cases pertain to confined portions of
space-time, they should not depend on our ability to completely circumnavigate the
space, something particularly obvious if the space in question is the universe as a
whole.

[2] It should strictly speaking be called "intrinsic curvature" but in general it is
simply referred to as curvature.

in higher dimensions to determine that the space is curved. Spaces can be curved in different ways in different directions and generically one needs more than one number to characterize the curvature.

5.3 Basic ideas of general relativity

Newton proposed a law of gravity that seemed to work well in practice but that left many questions unanswered. In particular, how does gravity work. How is it that a distant body like the Sun influences the Earth's motion, almost one hundred million miles away? This key concept of the action at a distance of a force, was finally tackled in the context of electromagnetism by Maxwell, almost two hundred years after Newton. The answer was to have a field. A charge influences the field, which in turn propagates that influence at a finite speed towards distant objects. This question of how does gravity exert its influence over space was one of the key ingredients motivating Einstein in his quest for the general theory of relativity.

In general relativity every object endowed with an energy induces curvature in space-time. The main set of equations of the general theory of relativity, known as Einstein's equations, schematically state that the geometry of space-time depends on the energy of matter. Space-time is not an empty vessel anymore where matter resides, but becomes a dynamical entity that is influenced by matter. In turn, space-time influences the motion of objects living within it. What Newton would have described as a particle acted upon by a gravitational force, Einstein describes as a particle living in a curved space-time. Particles that are far away from gravitating bodies behave as in a flat space-time and follow straight paths. Near a gravitating body the particles will follow the closest thing to a straight line in a curved space: they follow geodesics. The Sun bends space-time in the Solar system and as a consequence the Earth follows a trajectory that is a geodesic, it is an orbit rather than a straight line. The explanation of the motion of the Earth that general relativity provides does not require any mysterious instantaneous action at a distance of the Sun. The presence of the Sun deforms space in its surroundings and it is that deformation what affects the motion of the Earth. Obviously the Earth also deforms space in its surroundings. Since the mass of the Earth is smaller, the deformation is less pronounced and it is significant only at shorter distances, affecting nearby objects like the Moon.

Space-time therefore operates in an analogous way to the electromagnetic field when one has a moving charge. Oscillations of charges give rise to electromagnetic waves. Oscillations of masses like the motion of binary stars produce gravitational waves that also propagate at the speed of light. Gravitational waves are in many ways similar to electromagnetic waves. Masses in motion emit gravitational waves that could be detected by masses that operate as antennas and would oscillate under the passage of a gravitational wave. In fact, given that gravitation is universal and interacts with all forms of matter, gravitational waves induce a motion in any material object. Just like electromagnetic waves, gravitational waves can carry energy from the source to the antenna. Given that energy is one of the clearest evidences of the presence of matter, this type of result shows the extent to which we need to revise our preconceptions about space-time.

5.4   Conclusions

The existence of a maximum speed for interactions, the speed of light, as we have seen, leads, to the impossibility that certain kinds of events could influence others. If an interaction happens or not is something absolute, and this is a basic observation for the analysis of all elements that constitute reality. Events can therefore be ordered according to the possibility they have of interacting with each other, constituting what is known as the causal structure of space-time. Although neither space nor time have an absolute character anymore, the causal structure is absolute. Space-time therefore describes a network of events and their causal relations.

With the development of electromagnetism, a new actor appears on scene, the fields that propagate in Euclidean space as time passes. With the introduction of special relativity the precise distinction between space and time is lost and is replaced by that of space-time, associated with the causal order of events. But the most spectacular change occurs with general relativity. The very scenario in which actors moved, space-time in which particles and fields propagated, becomes an actor itself. All motion is relational and it describes how some actors move with respect to others. With the introduction of general relativity the development of classical physics ends. From the mechanics of

Galileo and Newton to the development of relativity the changes were enormous. As we shall see however, they pale in comparison with those that occurred with the development in quantum mechanics.

Summarizing, according to classical physics, there are three types of matter. 1) The material particles, which were the first to be recognized as physical objects. 2) The fields, which are able to transport energy and modify and be modified by the motion of charged particles. If we remember that light itself is composed of electromagnetic fields and that when an object hits another one the impact reduces to the interaction of the surface electromagnetic fields created by the bodies, we must conclude that we have direct evidence of all of them as good as that of particles. In everything we can see and touch, particles and fields work together. 3) Finally, space-time, since as we saw, not only is it affected by the different forms of matter and exchanges energy with them, but also provides the arena where the other physical objects live.

On the other hand one has to admit that the classical vision of matter is too abstract and does not give us much in terms of furthering our understanding about the real substance that physical objects, like a table, plants in a garden or ourselves, are made of. What can we learn from classical physics about a piece of iron? What can we say about its thermal, electrical or chemical properties starting from the fundamental laws of physics? Why do different material objects result from a finite set of elements and substances? Can we understand matter if ultimately we are unable to explain substances and chemical processes in terms of physics? Can we aspire to physics yielding a different vision of life, consciousness or man, if it has so little to say about a piece of iron? Quantum physics will allow us to respond to these questions and to take a great step towards a world view closer to the concrete objects of our experience.

Chapter 6

A first tour through the quantum world

6.1   Introduction

In previous chapters we have analyzed the worldview that results from classical physics. In spite of the transformations undergone by physics since the times of Newton and Galileo all the way to Einstein's, the classical world is nevertheless ruled by deterministic laws. What is happening right now will determine completely what will happen in the future. On the other hand it is taken for granted that there exists an objective reality "out there" that is not affected by our observations. The philosophical difficulties the worldview stemming from classical physics faces are not with interpretation, but with how to make compatible its predictions and the limitations of the classical theory with the richness of the physical world, including chemical phenomena, life, and our internal experiences as beings endowed with a conscience and responsibility. It does not seem natural to claim that our minds and bodies, which are physical objects, will not obey the same determinist laws as other material entities. Let us recall Voltaire's (1932) statement, which ironically attacked the attempts to escape this conclusion by saying: "It would be very singular that all nature and all the stars should obey eternal laws, and that there should be one animal five feet tall which, despite these laws, could always act as suited his caprice."

We will now start a long journey through the domains of the new quantum physics, as that will allow us to dig deeper in some of the major problems of philosophy with the hope of finding a coherent vision of the world and our own experience. Quantum physics has two fundamental properties from which very important consequences derive: it is quantum and it is probabilistic. First and foremost, *it is quantum; the very name of the new physics has to do with the fact that many fundamental quantities do not take continuous values, but certain preselected discrete values.* In classical physics the orbits of a negative charge orbiting around a positive one, as one may have in a hydrogen atom, can have any energy. Classical systems can lose energy gradually, for instance by emitting electromagnetic waves. In the quantum theory, the orbits of the atom can only have certain values for the energy. If the energy changes it does so in discrete values emitting fixed packets of electromagnetic waves called photons and jumping into a state of lower energy. To this fundamental property we owe the existence of matter organized in atoms, molecules and solid bodies. Every element, substance or crystal has a set of behaviors that completely characterize it. For instance a sodium atom will only emit light of a certain predetermined color, typically yellow. One can see this simply by sprinkling some salt on an open flame.

Secondly, *quantum physics is probabilistic.* A complete knowledge of the state of a system will not allow us in most cases to know with certainty its future behavior. For instance, if we know with complete precision the state of an electron, we cannot know its position in a subsequent instant. We can only assign it a probability that it will appear in a given region. The same holds if we try to predict its velocity or any other variable associated with its motion. A clear example of this are radioactive materials. They emit particles, for instance radioactive plutonium like the one used in nuclear reactors emits alpha particles (helium nuclei). We are not able to predict when a radioactive material will emit a particle. In the case of plutonium we know that half of the atoms would have emitted an alpha particle at the end of 24,000 years. This is known as half life. But we do not know when a particular atom will emit.

The status of probabilities in quantum physics is very different from the one in classical physics. In that case one uses probabilities to describe systems about which one has partial knowledge only. Such partial knowledge can always be supplemented by knowing more details about the process. For example, if one were to flip coins, there is an a priori probability that the flip will turn heads or tails. However, if one knows in great detail the characteristics of the coin and how it is being tossed, one could, with suitable computational power, predict if the next toss will be heads or tails. In quantum mechanics we have a great body of evidence indicating that the ignorance behind the probabilities if fundamental in nature. A more complete knowledge of the system is impossible: we just cannot predict where the electron will be five minutes from now. Even using the term ignorance seems inappropriate since as we will see, there is no evidence to indicate that the electron occupies a position in space or has a given velocity until it interacts with another system. For instance producing a dot on photographic paper or making a detector "click".

Many times there is the tendency to think that quantum phenomena are only relevant at microscopic scales and classical physics is enough to describe phenomena at ordinary scales. Such belief is incorrect both from a practical point of view as well as a philosophical one. In practical terms, quantum physics underlies many fundamental physical phenomena at ordinary scales: the stability of atomic and molecular structures, the very existence of solid matter and its electric, thermal and optical properties, the origin of chemical bonds and chemical reactions, the colors of certain substances, the elastic properties of materials, freezing and boiling points, properties of biological macromolecules and the possibility of coding genetic information are just some examples that rely crucially on quantum mechanics for their existence. A good fraction of the technological applications of physics of the last fifty years is based on quantum physics. Among them, lasers, semiconductors, superconductors, all of them pillars of the consumer electronic revolution that started fifty years ago and continues today and extends to computers, telecommunications and robotics.

Classical objects are composed of quantum objects, therefore, from an ontological point of view, their properties derive from collective behaviors of quantum objects. From the days of the original formulation of quantum mechanics up to today there has been significant progress in our understanding of the relationship between classical and quantum physics. However, even today we lack a completely quantum description of the world in objective terms. The central difficulty is to account for the world that is "out there" composed of unambiguous facts we all agree upon starting from the quantum world which consists of potentialities in constant evolution. The problems of interpretation of quantum mechanics are fundamentally rooted in that difficulty. In practical terms progress has been made through the use of the quantum theory by focusing on magnitudes that arise from certain measurement processes or observations of some classical properties of certain objects. Ultimately this is tantamount to presupposing the existence of a world accessible to our ordinary experience while renouncing the possibility of explaining its existence in terms of the quantum theory. This is very problematic and even contradictory with what the physicist does when she use quantum mechanics to study, for instance, what happens in the nuclear reactions in the Sun or when she describes the mutations due to the interaction of cosmic rays with the genetic material. In fact these processes that do not involve any direct measurement need to be described in a very unnatural way if you only relay on measurements or observations for making sense of the quantum phenomena. The interpretational problems are therefore basically linked to the lack of understanding of the privileged role that, as we will discuss, measurements have in quantum mechanics.

With the advent at the beginning of the 20th century of quantum theory, the mechanistic view, which considers the world as a mere mechanism whose reality is taken as unquestionable, cannot be sustained anymore. In its place arises another point of view, equally unsatisfactory philosophically, that claims that objective reality fades away and physics refers only to the results of our observations. While classical physics is difficult to reconcile with our subjective experiences, quantum mechanics, at least in its oldest and most widely considered interpretations, appears incompatible with an exterior world independent of our observations. Progress done in the last 50 years have led to glimpses of how this dilemma can be avoided and to have interpretations that account for objective reality and at the same time provide a new and richer point of view concerning matter.

In this chapter we will familiarize ourselves with quantum phenomena. Starting from empirical evidence and qualitative reasonings we will gain intuition about quantum behaviors. Many of these behaviors will appear counter-intuitive and we will have to adjust our commonly held beliefs to them. The task of updating our common sense requires two successive processes. The first task is a critique of old prejudices, which have to be acknowledged as such. The second is a building a new paradigm to understand reality. In this chapter we will start by accumulating evidence towards showing the limits of the classical description. We will analyze some simple quantum systems that will allow us to develop intuition about their principles. We will end by presenting in a simple way the axioms of quantum mechanics. They will allow us to appreciate the phenomena here described from a different perspective and acquire a firmer grasp of the theory.

6.1.1   Waves or corpuscles?

In quantum mechanics every elementary particle such an electron or the particles associated with light —the photons— have behaviors that recall the ones of classical particles and also wave-like behaviors. This duality expresses the inability of the classical concepts "particle" or "wave" to fully describe the behavior of quantum objects.

A given kind of quantum object will exhibit sometimes wave, sometimes particle, character, in depending on the physical setting. Einstein was the first to recognize that light has behaviors that can only be understood if one ascribes it corpuscular aspects. That is, if one thinks of it as composed of particles that carry an energy proportional to their frequency. The proportionality constant $\hbar$ is known as Planck's constant. Years later, de Broglie noted that conversely, in order to understand atomic behavior one had to think of the electrons as waves with a wavelength inversely proportional to their momentum.

Newton's mechanics describes particles that follow well defined trajectories. Microscopic objects are not particles in the classical sense, since they combine wave-like and corpuscular aspects. What is therefore a particle in quantum mechanics? This is a key issue to understand quantum systems. The fact that both light and matter behave as waves suggests that in the quantum theory the tendency to multiply the forms of matter can be reverted. In classical physics we had to add electromagnetic fields and later even space-time to corpuscular matter. Could it be that the observations of Einstein and de Broglie are the first step in a process of unification of matter forms thanks to quantum mechanics? Although unification is still not complete, since gravity has not been incorporated into the scheme in a completely satisfactory manner, the answer appears to be positive and the objective of this section is to begin to analyze its meaning.

The change in the description of the physical world imposed by quantum mechanics is so profound that we still have not absorbed all of its

implications even today. It questions the nature itself of what constitutes a physical fact. For instance, one needs to abandon the notion
that a particle is in a given position even when it is not being observed.
Furthermore, it is in general inconsistent to assume that a physical magnitude takes definite values in a physical system before a measurement
is made. In order to familiarize ourselves with quantum phenomena it
is good to consider the example of Young's double slit experiment but
now taking into account that the light is made of photons. We should
recall that what one observes when one interferes two beams in Young's
experiment is shown in figure 4.7 or 6.2. Interference makes light's intensity greater in some regions and smaller in others. If the light is
recorded in a photographic plate one observes very bright stripes and
dark stripes.

In order to study the corpuscular nature of light, that is, the behavior
of photons, one lowers the intensity of the beam in Young's experiment,
such that the source emits photons one by one. That is, emission takes
place in such a way that on average there is only one photon traveling
between the source and the plate. What happens in that case?

*It is observed that neither the predictions of the wave-like theory nor
those of the corpuscular one are true.*

Concerning the corpuscular theory, if one waits for a long time so
the impact of a large number of photons takes place on the photographic plate, one sees interference bands appear. Since the photons
pass through one at a time the bands cannot be the result of an interaction between corpuscles. Insofar the wave-like theory is concerned, if
the photographic plate is only exposed for a short time, allowing the
passage of a limited number of photons, one can see that each photon
leaves its own mark on the plate and therefore behaves like a particle.
What is happening is that the successive impacts of the photons occur
randomly but as the number of impacts grows the distribution becomes
approximately continuous and reproduces the interference pattern as
shown in figure 6.1.

As we saw when we discussed interference, the band pattern only
happens when both slits are open. The paradox that emerges is that
since the photons are being sent one by one, and presumably travel
through one slit, what could change if one of the slits is covered? The
experiment shows that the photons have two apparently contradictory
behaviors. Although each photon incident on the photographic plate
leaves a well localized dot as if it were a point particle that has impacted,

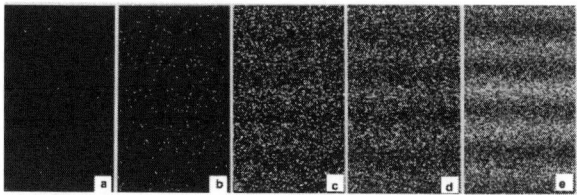

Figure 6.1: Snapshots of the double slit experiment showing how the accumulation of individual photons on the photographic plate gives rise to interference pattern typical of Young's experiment. Image: Akira Tonomura.

the distribution of such dots reproduces a pattern typical of waves.

The distribution of intensities in the interference pattern is different from the one that would be obtained by adding the intensities resulting of going through each slit individually. This shows that the photon is not behaving as a particle. The interference pattern can only be accounted for by assuming that each photon interferes with itself and goes simultaneously through both slits. Somehow the photon exhibits a wave-like behavior when it propagates and only behaves as a particle when one attempts to establish its position, as it happens in the photographic plate. The physical events, such as the appearance of a dot on the photographic plate only occur in a measurement process. The same behavior is observed if the double slit experiment is carried out with electrons instead of photons, as de Broglie had predicted.

In the case of electrons if one attempts to measure through which slit they pass, say, by illuminating it, the observed distribution of intensities will look like the one indicated by "C" in the figure, and we will observe a distribution that corresponds to classical-like particles going through one of the slits. The interference pattern associated with the wave-like motion disappears. We therefore see that in measuring the position the wave-like behavior is destroyed and the system behaves in a corpuscular fashion. It is important to note that what we refer to as measurement process does not necessarily require the presence of a human experimental physicist, it only requires that the microscopic system (in the example, the electron) interact with another system whose behavior is approximately classical (the photographic plate, or an intense enough

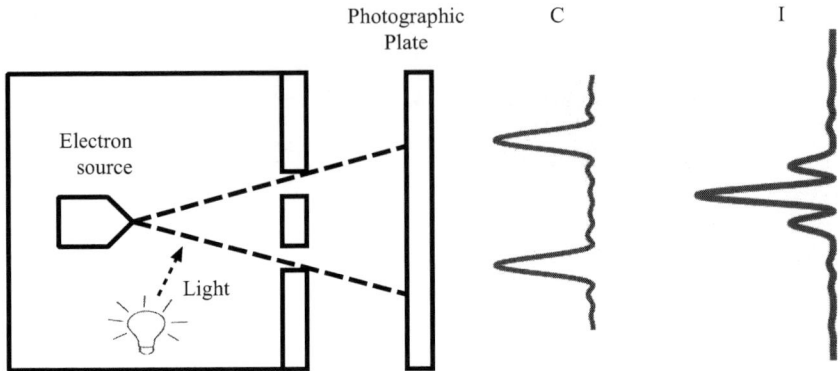

Figure 6.2: The double slit experiment. Attempting to observe the electron before it moves through the slits destroys interference.

beam of light to intercept the electrons).

We have therefore found that when one carries out a measurement on a microscopic system (in the example measuring a position) it get fundamentally perturbed (in the example the interference pattern disappears). In classical systems this situation does not arise: the effect of the measurement apparatus can be made in principle as small as one desires. If in the double slit example one uses a beam of light of low intensity, some electrons will make it through and the interference pattern will reappear. What cannot happen is to both determine through which slit the electron passed and to have the interference pattern. The idea of a classical trajectory does not work in this example, the individual electrons behave as waves as they propagate and only have a position if one interacts with them. We will later return to this idea, we will see that the natural way to understand this is to abandon the language of particles and waves and to talk in terms of events.

Concerning the distribution of photons —or electrons— on the screen, it is impossible to determine where a specific photon will impact the photographic plate. Even if we had complete information about the initial state of the beam of light hitting the screen with the slits, one will only be able to assert that there is a certain probability that the photon will land on a specific region of the photographic plate.

Summarizing: a) quantum particles do not behave either as waves or particles; b) their behavior under a measurement of position can only be predicted probabilistically; c) the distributions of dots reconstructs, when a large number of dots is deposited on the screen, the bright and dark striped pattern observed in the case of the electromagnetic waves. Therefore the value of the probability is given by the intensity of the interference pattern, in the case of photons it is proportional to the square of the electric field.

6.1.2 The uncertainty principle

Let us return to the double slit experiment. As long as we do not force the electron to answer the question of what is its position, the electron does not choose any answer. It behaves as if it went through both slits. When we attempt to determine through which slit it went, by measuring its position, we radically alter its behavior. We will now know precisely where it goes through the screen, but we would have lost all information about its momentum. That modification of its momentum alters the distribution of the electrons on the photographic plate, making the interference pattern disappear. Recall that knowing the momentum of an electron amounts to knowing its velocity. Unlike in classical mechanics, the system always retains a certain level of freedom independently of its behavior. That freedom is embodied in the uncertainty principle. The limitation in the simultaneous measurement of position and momentum is one of the possible uncertainty relations. Another important one is that relating the uncertainty in the energy and the time a system takes to change significantly its behavior. In quantum mechanics a system whose energy is known precisely does not evolve in time. Reciprocally, the energy of a system can change importantly in small intervals of time.

We can only know with certainty that a quantum mechanical system that is measured will give an answer chosen from a possible set of answers and to compute the probability of each answer. One needs to think about quantum mechanical systems as capable of having certain behaviors that are actually realized when a measurement takes place and give an answer that is the result of the measurement.

Anastopoulos in his recent book (2008) on the origin of the quantum theory observes that "For Heisenberg and Bohr, the uncertainty principle was a vindication of their overall philosophical perspective. No

matter what the precise formulation of quantum theory is, the wave
particle duality alone shows that it is impossible to measure with full
accuracy what Newtonian mechanics characterized as the fundamental
physics quantities of a particle's motion. If one accepts that physics
ought to be formulated only in terms of what is actually measurable,
the notions of physical quantities been represented by numbers loses
any meaning because position and momentum are not measurable simultaneously. "In the strong formulation of the causal law 'If we know
exactly the present we can predict the future' it is not the conclusion
but rather the premise that is false" Heisenberg wrote "We cannot know
as a matter of principle the present in all its details.""

This observation will lead to a new way to describe the information
one has available about a system in a given instant of time: the quantum
state. This will not be described as in classical physics in terms of
the position and momentum of the particle but by a new object that
Schrödinger called the wavefunction and later would be renamed state
vector.

6.1.3   Wavefunctions and wave mechanics

De Broglie had advanced the daring hypothesis that if light had an
associated particle behavior (photons), by symmetry, particles should
have a wave behavior. His hypothesis gave a nice underpinning to Bohr's
prescription for the admissible energy levels of the atom. Just like the
strings of a violin can only vibrate in a finite set of frequencies, the waves
associated with the electrons could only live in closed atomic orbits that
could fit an integer number of wavelengths.

If particles behave as waves, what kind of equation do they obey?
For photons we know the answer, it is the equation of electromagnetic
waves. What is the equation for the waves associated with the electrons
in atoms? Schrödinger discovered his equation imposing that the wave,
represented by a function that is usually denoted by the Greek letter Ψ
and that takes a value at each point x in space $\Psi(x)$ and it propagates
in such a way that it reproduces in some cases the motion of particles
as described by Newtonian mechanics. He noted that the wave equation
for the photon reflected the relation between energy and momentum for
photons $E = c\,p$ and he postulated that the equation for matter waves
has to reproduce the relation between energy and momentum of classical
mechanics $E = p^2/(2m)$.

Studying with Schrödinger's equation the solutions that represented electrons in stable orbits that remained bound to the nucleus, he noted that such solutions only existed for values of the energy that coincided with the permitted energies of Bohr's atomic model. They reproduced the spectrum of emission lines that spectroscopists measured. This experimental confirmation gave great hopes of understanding quantum phenomena purely in wave terms. This expectation, sparked by findings of Schrödinger, Einstein, Planck, ended up not panning out. The first opponent was Heisenberg whose vision of quantum mechanics centered on the quantum jumps associated to the energy levels that seemed to be fundamentally discontinuous and random.

Another problem of Schrödinger's equation that complicated the wave interpretation is that Ψ is not a real function, but takes values in the complex numbers. Whereas real numbers are associated with physical magnitudes, complex numbers, which include the square roots of negative numbers cannot be directly associated with observable quantities. What exactly is therefore described by Ψ?

Physicist Max Born, who was Heisenberg's adviser, would be the one to give a definitive answer to that question, a few yeas after Schrödinger's equation had been introduced. With it, he would destroy the hopes of those who wanted to reduce quantum phenomena to wave phenomena. Born's observation is that according to the uncertainty principle it is not possible to get precise values for the magnitudes of a system. When an experiment is repeated with the system in the same state, one obtains results for the measurements that are not identical. Since one cannot get precise predictions one can only aspire to get probabilistic predictions. Just like the intensity of light is proportional to the square of the electric and magnetic fields, Born postulated that that the probability of finding a particle in a given point x is proportional to the square of $\Psi(x)$. Since the latter is complex, more precisely the probability is proportional to the absolute value of $\Psi(x)$ squared. [1] This assignment of probabilities is known as Born's rule. Strictly speaking, the probability of finding something *exactly* at a point is actually zero, what one needs is the probability of finding a particle in a small region surrounding a

---

[1] Real numbers can be represented by the points of an oriented straight line, more specifically by the distance from the point in question to the origin. Complex numbers are represented by points in a plane. The absolute value of a complex number represents the distance to the origin of that plane, the latter being the zero complex number.

point, that is proportional to the absolute value of $\Psi(x)$ squared times the volume of the small region (this also explains why it is zero if one asks for the probability at exactly a given point).

Moreover, the equation for $\Psi(x)$ is linear, meaning that if one has two solutions to it, one can superpose them to form another solution. This superposition property is one of key behaviors of quantum mechanics we will study now.

6.2   The fundamental rules of quantum mechanics

Given that the main goal of this book is to analyze the implications of quantum mechanics for our understanding of the world, we need to introduce the theory in somewhat more detail than what we did for the classical theories. The basic rules of quantum mechanics are summarized in a set of axioms that require a certain level of mathematical sophistication to introduce. So we will not introduce the theory in that way. We will proceed by presenting its application in two very simple systems: a set of polarized photons and particles with spin. Studying these systems will allow us to exhibit the typical behavior of quantum systems and outline the fundamental laws that rule them. Remarkably, a good fraction of the revolutionary or controversial aspects of quantum mechanics can be exhibited through examples of polarized photons or electrons and other massive particles with spin, as we shall see.

6.2.1   Polarized photons

Let us start by recalling the meaning of polarization of electromagnetic waves. Initially, we are talking of classical electromagnetism only. When we studied electromagnetic waves, we referred to linearly polarized waves. We saw that the electric field oscillates in a plane, whereas the magnetic field oscillates in a perpendicular plane to the one in which the electric field oscillates. We adopt the convention of calling the electric field plane the *polarization plane*. In the chapter on classical electromagnetism we studied how two waves with the same plane or polarization that propagate in the same region of space behave. We saw that we had phenomena of constructive and destructive interference.

Ordinary sources of light, like a lightbulb do not produce light of a single frequency nor with a given polarization. The light is produced by individual molecules that are moving violently in the filament of the lightbulb due to the high temperature of it. The resulting electromagnetic field is the superposition of the fields produced by each molecule and the latter are randomly polarized. The resulting light can be easily polarized using a filter, that allows the passage of light of a given polarization. Filters of this sort some times are used in sunglasses. From now on we will consider that we are working with polarized light. Let us assume that the polarization plane forms an angle θ with respect to the vertical. If such light hits a polarizing filter oriented along the vertical line, one observes that some light makes it through. If one now rotates the polarizer one observes that the intensity diminishes as the angle between the polarization plane and the polarizer tends towards ninety degrees, and no light makes it through when both planes are perpendicular. The amount of light that makes it through when the angle is not ninety degrees is given by what is known as Malus' law which states that the fraction of intensity that makes it through goes as the cosine of the angle between the polarizer and the polarization plane, squared.

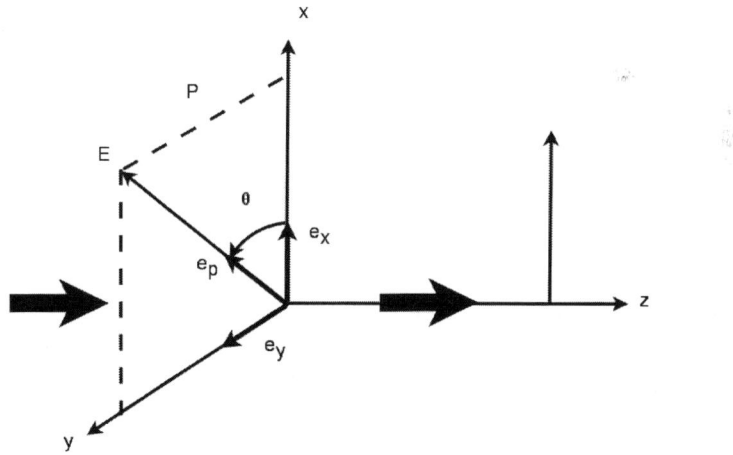

Figure 6.3: The orientation of the polarizer and the field.

Up to here we have concerned ourselves with classical electromagnetic waves. Quantum behavior will be observed if we lower the intensity of the incident light so photons arrive individually to the polarizer. If the incident beam is monochromatic the photons will have an energy $h\nu$. One observes that the photons that make it through the polarizer keep the same frequency and energy. Every time a photon hits the polarizer it either makes it through or it is absorbed. One cannot have only a fraction of the photon pass through. It is impossible to know what an individual photon will do, one only knows they will make it through the polarizer with a probability that goes as the cosine of the angle squared. If one sends a large number of photons, a fraction equal to the cosine squared will make it through, therefore reproducing Malus' law.

From this experience we can extract the following lessons that generically characterize quantum behavior (Cohen *et al.* 2006).

i) All the information we have about the behavior of the photons is contained in their state: in this case we can characterize the state through the polarization plane of the photon. So it is given by a unit length vector in a two dimensional plane perpendicular to the direction of propagation. In the figure the photon propagates along the z direction and the polarization is in the $x - y$ plane.

ii) The detector can only give some privileged results: we call them *proper results*. In the experiment described only two results are possible: the photon made it through the polarizer or it did not. One says that the results of the measurement are quantized, there exists a discrete set of results, the photon made it or not. This should be contrasted with a continuous of results for the classical case, depending on the angle of the polarizer.

iii) To each result one has associated a given state, *called proper state*. The proper state associated with the photon that makes it through is the unit vector $\mathbf{e}_x$ (see figure) and the one to the one that does not make it is $\mathbf{e}_y$. If the photon is prepared in the $\mathbf{e}_x$ state it will make it for sure, and if one prepares it in the state $\mathbf{e}_y$ it will not make it for sure. The idea is that if the system is in a proper state prior to the measurement then one will obtain the proper result with certainty.

iv) *Born rule:* The probability of obtaining a proper result is given if one knows the initial state. One simply decomposes the initial state $\mathbf{e}_P$ as a linear combination of the various possible proper states. The coefficient of the proper state of interest squared gives the probability that the corresponding proper result will be obtained.

v) When the light makes it through the polarizer its polarization changes and it emerges polarized along $\mathbf{e}_x$. As a consequence, if the state of the photon was initially characterized by a polarization $\mathbf{e}_P$ the final state is $\mathbf{e}_x$ irrespective of $\mathbf{e}_P$. There is therefore a sudden change of the state as a consequence of the process of measuring the polarization. The state of the system suffers an uncontrollable modification in the measurement process.

A central concept in any quantum description is the idea of *super-position*. An individual photon polarized along $\mathbf{e}_P$ is in a state that can be considered like the superposition of proper states that would yield different classical behaviors, for example $\mathbf{e}_x$ and $\mathbf{e}_y$. When the photon is in such a state it can choose among classically excluded behaviors, in the example to make it or not through the polarizer. Notice that a given state can be decomposed in many ways. For instance if we rotate the axes x, y to x', y' and we are interested in the behavior of a polarizer along $\mathbf{e}_{x'}$ then it would be convenient to decompose the state as a superposition of states $\mathbf{e}_{x'}$ and $\mathbf{e}_{y'}$.

It will be convenient to use a different notation to denote the states of the systems that was initially invented by Dirac. If we are talking about a proper state with polarization along the x axis instead of $\mathbf{e}_x$ we will denote the vector $|e_x\rangle$ and if oriented along y $|e_y\rangle$. The symbol $|\rangle$ indicates that it is a possible state vector of the system and the label indicates the specific vector that we are referring to.

The previous analysis illustrates five of the six fundamental axioms that rule quantum systems. We still have to refer to the axiom that establishes how quantum states evolve. We will discuss it in the next sections. Notice that to analyze the polarized photon experiment we only needed to assume that for a large number of photons we recover the classical result and that light has associated with it quanta of energy that cannot be divided.

The preceding rules tell us in what polarization state the photon emerges after the measurement. If it goes through a polarizer aligned with the x axis it ends up in the state $|e_x\rangle$. It does not matter what the state was prior to the measurement, it suffers a sudden change and ends up in $|e_x\rangle$. As a consequence one cannot, with a single measurement, determine the state of the photon prior to the measurement, since the latter destroys the original state and does not yield enough information to reconstruct it. Only if we have a large number of photons in the same state, and we measure them, we could, via a statistical analysis of the results, determine the initial state of the photons.

6.2.2   Spinning particles

Most elementary particles like the electron have a property that is analogous to the polarization of photons, called spin. One can intuitively think that the electrons behave like a classical spinning top. The latter has a property named angular momentum. However in quantum phenomena there is an important difference and it is that the value of the angular momentum is quantized with respect to magnitude and orientation. If we wish to think of electrons like little tops, their axis cannot be arbitrary and the rotation speed neither.

In the case of electrons, protons or neutrons, the angular momentum (spin) is $s = \sqrt{3}\hbar/2$ where $\hbar = h/(2\pi)$ and, surprisingly, in the quantum theory the projection of the angular momentum on any axis can only take two values. For instance, the projection on a vertical axis can only be $S_z = \pm\hbar/2$. Again, surprisingly, the axis of rotation of an electron or other particles with this value of spin (for conventional reasons one calls them "spin 1/2 particles") is never "perfectly aligned" with a given direction, since its projection on any direction is always smaller than the value of the angular momentum. It is like the spin precesses with its axis rotating around the direction of interest. One calls the positive component of the projection conventionally "spin up" and the negative one "spin down". Just like in the case of photons, we have two proper states. We will use the notation $|z\text{up}\rangle$ for an electron with component along the z axis up and $|z\text{down}\rangle$ for the spin down component. If we further the analogy with the top, since electrons have an electric charge, one should consider a charged top. The charges on the top rotate with it and we know that should generate a magnetic field. That makes the electrons behave like little compass needles that point in the same direction as the spin. A similar situation is true for protons. More surprising is that neutrons, which do not have electric charge, have a similar magnetic behavior. The origin of that behavior of neutrons is that they are composite particles. They are composed of quarks, which are charged, and what is true for electrons is also true for quarks. Although the net charge of the quarks forming a neutron is zero, they still give rise to magnetic fields. Another example of particles that are neutral but have magnetic fields are silver atoms. These were of great historic importance because experiments with them led Uhlenbeck and Goudsmit to the discovery of spin. The magnetic properties of particles with spin will allow their manipulation in experiments similar to those

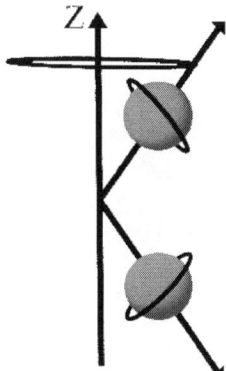

Figure 6.4: The axis of spin precesses around the z axis. The top diagram corresponds to spin up, the bottom to spin down.

we carried out with polarized photons, as we will see.

The analogy between the electron spin and the polarization of photons is not accidental. Indeed, photons have a basic form of spin, basically the photon associated with electromagnetic waves that are circularly polarized are particles of spin $\hbar$ with the axis of rotation aligned along the direction of propagation. The spin of photons is known as *helicity*, it is positive when the axis of rotation is aligned with the direction of propagation and negative when it has the opposite alignment.

The *Stern–Gerlach experiment* (figure 6.5) exploits the magnetic properties of spinning particles. The main element is a magnet that creates a magnetic field between its poles that is more intense the closer one gets to the top pole. In the center the field points in the z direction as the figure shows. That field exerts on a silver atom injected into the region a force that would deflect the motion depending on the z component of the spin. Atoms injected with spin up would move upwards and those with spin down downwards. Although their trajectories would not be precisely defined due to the uncertainty principle, they would lump together in two well differentiated regions. If one had done the experiment with classical particles, one would have got a continuous distribution depending on the initial orientation of the spin as they fly into the apparatus, as indicated in the figure. As in the example of the polarized photons, the experiment indicates that silver atoms have a quantized behavior with two possible values for its spin.

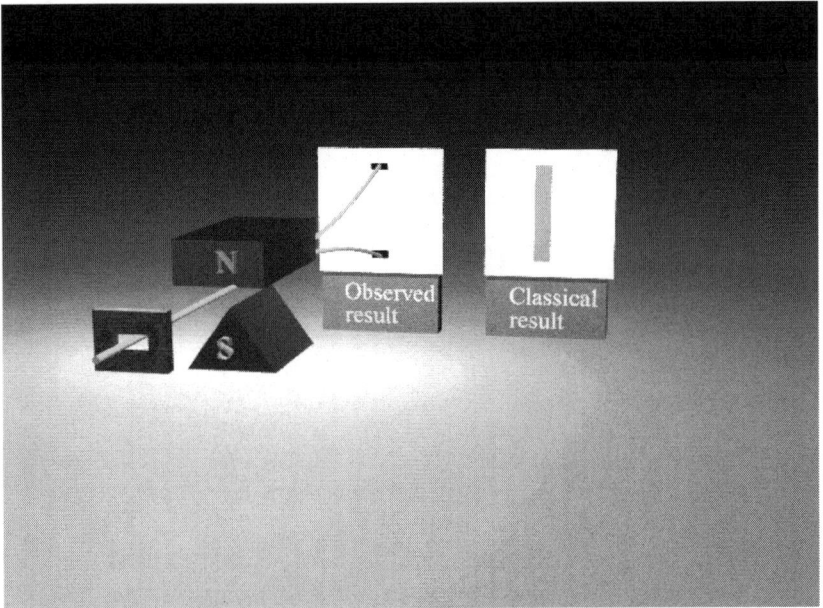

Figure 6.5: The Stern–Gerlach experiment.The pointed shape of the bottom magnet makes the magnetic field inhomogeneous in the z direction. That is needed in order to deflect the silver atoms due to how they couple to the magnetic field.

To analyze the experiment in more detail we need to know the initial state of the silver atoms. Let us assume that we use the Stern–Gerlach apparatus of the figure to prepare the initial state of the silver atoms, for instance, by blocking the passage of the spin down atoms. We can then take a second Stern–Gerlach apparatus forming an angle with respect to the first one and we would be in a situation similar to the one we set up with the polarizers for the photons, reaching similar conclusions.

This set of experiments can also be summarized in five conclusions

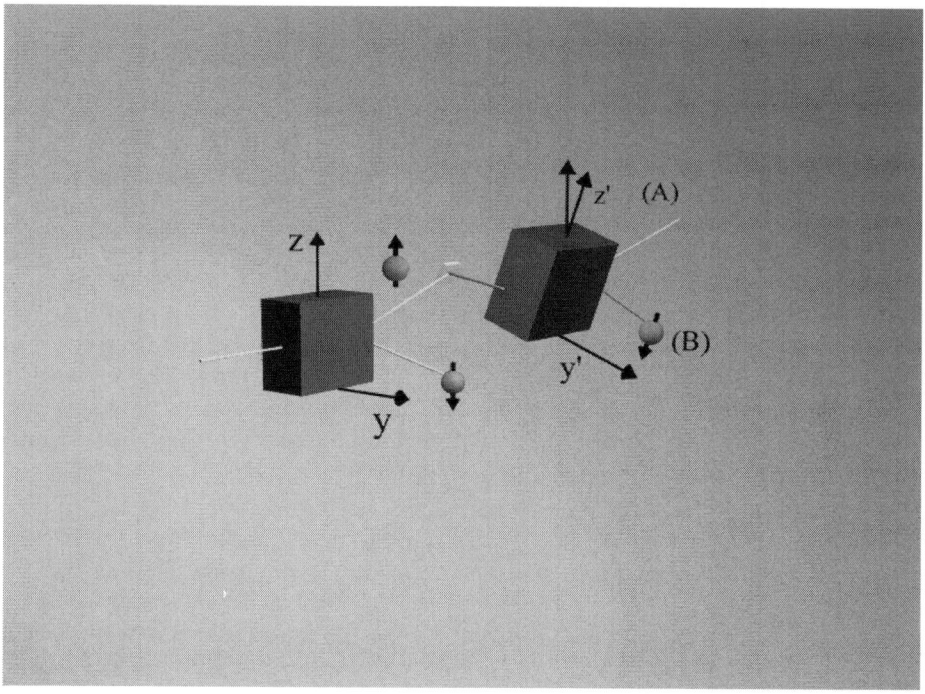

Figure 6.6: Using a Stern–Gerlach device to prepare a polarized state to go through a second Stern–Gerlach device.

as we did for the polarized photons.

i) All the information that we have about the behavior of the silver atoms is contained in their state: in this case we can associate the state as prepared by the first Stern–Gerlach apparatus. For that purpose we illuminate the region close to the branch two of the figure. If one detects the passage of an atom through that branch, the state of the silver atoms previous to the second apparatus will be $|z \text{ up}\rangle$.

ii) If one illuminates the beams of particles after passage through the second Stern–Gerlach apparatus, only certain privileged results will be observed: we call them proper results. In the described experiment there are only two of them: one detects the silver atom either in the upper part of the photographic plate facing beam A and has spin in the direction z' or it is detected in the portion in front of B and has spin down in the direction z'. Each physical property has its corresponding set of proper results. Other properties, for instance different components of the spin will be measured by different measuring apparata, in the example considered Stern–Gerlach apparata with different orientations of the magnetic field.

iii) To each proper result corresponds a proper state. The one associated with the silver atoms that emerge in the direction (A) is $|z'\text{up}\rangle$ and those that emerge in the direction (B) are $|z'\text{down}\rangle$. If we align both devices such that z coincides with z' then the electrons will be reflected upwards by the first apparatus, which were in the state $|z'\text{up}\rangle$ and will surely pass through the second apparatus deviating upwards in the direction z'. The idea is that if the system is already in a proper state before the measurement then one will obtain the corresponding proper result with certainty.

iv) Born's rule: The probability to obtain a proper state is given if one knows the initial state. To calculate it one decomposes the state as as a sum of the various proper states times coefficients. The probability of obtaining a given proper state is proportional to the square of the coefficient that appears in front of it in the linear combination (since in general the coefficients are complex numbers it is really proportional to the modulus square).[2]

v) When we detect the silver atom passing through (A) after leaving the second Stern–Gerlach device, we must conclude that the atoms have suffered a change in their state. They enter in the $|z\,\text{up}\rangle$ state and emerge in the $|z'\,\text{up}\rangle$ state. As a consequence there is a sudden change of the state due to the measurement process, that is, due to the interaction with the detector that determines that the atom passed through (A). The state of the system changes uncontrollably during the measurement process.

The interaction processes with detectors play a key role in the phe-

---

[2] A linear combination would be of the form $|z\,\text{up}\rangle = a|z'\,\text{up}\rangle + b|z'\,\text{down}\rangle$. The coefficients a, b are complex numbers. The latter include the real numbers as a particular case and can be represented by pairs of real numbers $a = (a_r, a_i)$. The modulus is given by the square root of the sum of the square of the of two real numbers and is denoted with vertical bars, i.e. $|a| = \sqrt{a_r^2 + a_i^2}$. In the above superposition the probability of measuring z' up is $|a|^2$. One can always adjust the length of the state vectors such that $|a|^2 + |b|^2 = 1$ and that means that the only two possibilities are up or down.

nomena we have just described. If such interactions do not take place, on output from the second Stern–Gerlach device the state will still be $|z\,\text{up}\rangle$, something that can be verified rejoining the beams (A) and (B) into a single one and measuring it with another Stern–Gerlach device identical to the first one. The end result of these experiments reinforces the idea that it is incorrect to think that the silver atoms choose to go through either branch (4) or (5) before being measured. Such choice never takes place since if one recombines the beams without making a measurement the components $|z'\,\text{up}\rangle$ and $|z'\,\text{down}\rangle$ recombine giving the original $|z\,\text{up}\rangle$. These processes in which no measurement takes places are related to another fascinating quantum phenomenon known as "delayed choice" but we will not go into details here.

It should be noted that in the previous discussion we have not considered the full quantum state of the particles. We have only included in the states complete information about their spins and left out all information about their spatial position. Such information would have been relevant, for instance, in the two slit experiment, but not in the Stern–Gerlach one which only depends on one projection of the spin. To discuss the spatial behavior of a quantum state one needs to introduce proper states associated to finding the particle at a given point in space. For instance $|x\rangle$ is the state that corresponds to finding the particle in the point x. To compute the probability of finding the particle at x one must start by decomposing the quantum state of the particle, let's call it $|\psi\rangle$ into a sum of proper states $|x\rangle$ evaluated at all points in space. Since the variable x is continuous the sum becomes what is known in mathematics as an integral. There will be one coefficient in the sum per point in space x so we denote the coefficient by a function $\psi(x)$. The probability of finding the particle in a small volume V_x around a point x is given by $|\psi(x)|^2 V_x$.

6.2.3   The probabilistic nature of quantum mechanics

Irrespective of how surprising the rules of quantum mechanics may initially appear to be, they are quite natural when one accepts the quantum (discrete) nature of microscopic objects. In the 1927 Solvay conference Werner Heisenberg and Max Born, the fathers of the probabilistic interpretation of quantum states, argued that "quantum mechanics is a complete theory" ... "Its basic physical and mathematical hypotheses are not further susceptible to modifications". Such a strong statement may appear surprising, but it was supported by very solid theoretical and experimental evidence. That point of view continues to be the dominant today and basically it rejects any possibility of explaining through some underlying mechanism the probabilistic predictions of the theory. It should be noted that there was no unanimity at the time and Einstein and Schrödinger never completely accepted that point of view.

As Giancarlo Ghirardi (2007) observes "...we can assert that wherever the quantum description of physical systems is assumed to be valid and complete, the quantum probabilities, in the language of the philosophy of science, are non-epistemic, which means that they cannot be attributed to ignorance, or to a certain gap in information about the system, which, if we could only fill the gap, would transform probabilistic assertions into certain ones."

Microscopic phenomena are therefore considered *fundamentally* random, contrary to what happens with most processes described by probability theory, which are fundamentally deterministic. Probabilities like those that describe processes like flipping a coin or rolling dice are known as *epistemic probabilities*. In those cases we need to apply probability theory because we are ignorant about the initial conditions in which the coin is tossed or the dice is rolled. If we knew them precisely we could use Newton's laws to figure out the result of the toss or roll. Since both the dynamics of the coin or dice depend very sensitively on the initial conditions it is in practice very difficult without some sophisticated equipment (and a lot of computational power), to predict the result of the toss or roll.

The case of quantum mechanics is different. Randomness of results is inherent to the formalism which only allows to assign probabilities to events associated with the different proper results. There is nothing else we can measure, no sophisticated equipment or computational power, no additional laws that we can use to determine the end result with certainty. One can only have results with probability one if the initial state is a proper state of the measurement device.

6.2.4   The basic concepts of quantum mechanics; systems, states and events

Let us try to generalize the lessons learned in the experiments that have just been analyzed to an arbitrary quantum system. We will follow

Ghirardi in this section. The goal of any physical theory is to predict, knowing the initial behavior of a system and the laws that rule its dynamics, its future behavior.

Let us first consider a Newtonian mechanical system to explore similarities and differences. The previously alluded to predictive process has three stages:

i) *Determination of the initial state*, which in classical physics implies to measure positions and velocities of all particles in the system.

ii) *Study of the evolution of the state of the system*. Given the initial conditions and knowing the details of the interaction forces of the particles among themselves and with the environment, the system evolves according to known mathematical equations.

iii) *The prediction of future behavior using Newton's laws.*

In a quantum system the analogue of these stages are as follows:

i) *Determination of the initial state*. We make measurements to extract the needed information to determine the state vector unambiguously. Following Schrödinger we will use the notation $|\psi, 0\rangle$ to designate the initial state vector (at time $t = 0$). The measurement process is constrained by the uncertainty principle. In other words the complete determination of the quantum state does not imply that al the magnitudes associated with this state will take definite values. As a consequence of this we will never have enough information to determine the position and velocity of a quantum particle with arbitrary precision. If one knows the position with high precision there will be a large uncertainty in the velocities that the system can take. In the case of the spinning particle it has a definite component on the z axis but its components along the x or y axis cannot be determined simultaneously.

ii) *Study of the evolution of the system*. As in the classical case we wish to determine how the system evolves from a state at $t = 0$ to some final time t if one knows the forces acting on the system. Schrödinger's equation does that job (we will not discuss its form in detail as it is not relevant to understand the basic principles). The equation is deterministic, that is, given a state at $t = 0$ $|\psi, 0\rangle$, the state at some later instant t is uniquely determined $|\psi, t\rangle$ by solving the equation. Moreover the equation is linear. That means that if the system is in a state that is a sum of states at $t = 0$, it will be given at t by evolving each state in the sum and summing them. This would allow us, for instance, to compute in detail how the initial state $|z\text{up}\rangle$ in the Stern–Gerlach experiment evolves while going through the magnetic field into a superposition $|z'\text{up}\rangle$ and $|z'\text{down}\rangle$. The deterministic evolution of systems while they are not measured is one of the central properties of quantum mechanics. This evolution is called unitary and it ensures that the probabilistic interpretation of quantum mechanics is preserved in time. Basically if the state of a system evolves unitarily, all the information about the system is preserved.

iii) *Prediction of future behavior.* The probabilistic nature only arises when the system interacts with measuring devices to yield the desired measurements. The measurements are associated with *events* in the measuring devices that characterize the different proper results that can arise from the measurement being made. If we are measuring the components of the spin along a given direction z' the proper results, as we saw, will be up or down along z'. In each case either the detector located in the region (A) will "click" or the one in region (B) will. The probability of each event is given by the square of the coefficient of the relevant proper state in the decomposition of the state in terms of a sum of proper states. The role of "clicks" in this example, or what is more generally known as *events* in measurement devices is key in quantum measurements. To begin with, the latter are associated with the interruption of the deterministic evolution given by Schrödinger's equation and they transform sums of states into states associated with one of the terms of the sum. That is, the state will suffer an abrupt change passing the initial superpositions of up and down states to the final state that will be either $|z'\text{ up}\rangle$ of $|z'\text{ down}\rangle$.

The postulate that the state changes in such a way in the measurement process when an event takes place is known as *reduction postulate.* The probabilistic predictions are about outcomes that result from an event in the measurement process. It is the only thing the physicist has access to. The quantum jumps and the fundamental probabilistic nature of the theory are associated with the occurrence of events like those that happen in measurement devices.

It should not be surprising that given that quantum mechanics not only discusses observable phenomena but a world of potentialities that exists prior to the occurrence of events, the type of math one really needs to fully describe the quantum theory is somewhat elaborate. We have only mentioned some aspects of it that are easily describable, in part by considering very simple systems as examples. For instance, in classical mechanics one uses a real number to describe the position of a particle. In quantum mechanics to describe the same quantity one uses something sometimes called "q-number" which can be represented as an array of numbers that basically contains information about the proper states that can arise from measuring that quantity. Another manifestation of the greater complexity of the math of quantum mechanics is that the coefficients that represent the components of a state vector are complex numbers, a natural generalization of the real numbers that describe physical measurements. A good fraction of the interpretational problems of quantum mechanics arise as a consequence of the abstract nature of the fundamental mathematical objects of the formalism: states and properties.

6.2.5 Conclusion

Quantum mechanics does not refer to a world we observe but to a world of potentialities. It describes what can happen, always in terms of probabilities. It tells us how physical systems go from probabilities to observed facts. It indicates how and in which circumstances, for instance when we measure a system, the latter chooses one of the possible results and one goes from a world of potentialities to one in which one of the possible options is observed. The states of a system characterize the disposition that the latter has to behave in a certain way when measured. Such disposition is always probabilistic in nature. One can only know what is the disposition of a system to give a certain result with a given probability.

We need to emphasize once more that the probabilities that we discuss are always probabilities of measurements or results or events that occur in the measurement process. Quantum mechanics, at least in the traditional version that we attempted to introduce here, is a theoretical framework that only allows statistical statements about possible results in measurement processes. If the predictions of quantum mechanics are only valid for what is observed in a measurement process or are valid in general for any event is something that needs to be elucidated and that is not decided in the traditional presentation due to Bohr, later known as the Copenhagen interpretation. We will discuss this issue in the following chapter.

Chapter 7

A surprise and a mystery of quantum mechanics: entangled systems and measurements

Anybody who is not shocked by quantum theory has not understood it.
Niels Bohr

This chapter will play a key role in the ideas we wish to present in this book. Indeed, as if the ideas reviewed in the last chapter about quantum mechanics were not fascinating enough, there are two key aspects we have not discussed yet. For those of us who believe that quantum mechanics provides a description of the real world, they are quite surprising and to a certain extent mysterious. We are talking about the non-locality properties of quantum systems and of the issue of measurement. Incorporating them into a realistic interpretation will open the door, as we shall see, to a completely novel way of conceiving reality. A new way of thinking that appears capable of harmonizing the scientific view of the world with a philosophical tradition that conceives humans as free responsible beings capable of behaving rationally in a world full of meaning.

Science is not compatible with mysteries. In one way or another they must be eliminated. If they cannot be explained away they must be shown to be non existent. It is therefore not surprising that in the face of these apparently mysterious properties of quantum mechanics the first attitude was to deny their existence. In this view physics did not have a role in understanding reality and an instrumentalist position was taken. As long as physics could make predictions verifiable by experiment all was well, no matter what the ultimate meaning of the theory in question. Bohr was the first in defending a position of this sort in order to make sense of quantum mechanics. Although both Bohr and Heisenberg seem to have been influenced by the Vienna Circle and the positivist philosophers, they never showed an explicit support for their position. As physicists, they were essentially adherents to instrumentalism. This originates in the observation that physics, strictly speaking, is limited to accounting for results obtained in experiments, and therefore it is excessively ambitious to think that scientific knowledge will allow us access to a reality independent of observations. It is not assumed that our knowledge is constrained by a priori limitations of our rational ability or that, as Kant believed, there exist notions like causality or space-time without which all knowledge is forbidden. It is simply stated that physical theories must be considered instruments that allow to predict, starting from some observed facts, the results of future observations. The goal of the physical sciences, according to this point of view, is none other than to account for communicable experiences. There is wide agreement among physicists, be they realists or not, that this is the basic goal of physics. The differences arise as to how sufficient this goal is. Whereas for an instrumentalist atoms and molecules are formal instruments that are useful to infer from them the behavior of experiments directly accessible to our experience, for the realists they are the "stuff" of which reality is made.

A factor that no doubt has contributed to the predominance of the instrumentalist point of view is the failure for many years of attempts to reformulate quantum mechanics using the realist language of pre-1920's physics. The point of view of Bohr and Heisenberg, later called the Copenhagen interpretation, became so preeminent in the first half of the 20th century that it was considered a waste of time to discuss alternatives. Jeremy Bernstein (2011) wrote a great book about the creators of quantum theory that includes the following anecdote that occurred in 1957 at the Institute for Advanced Studies in Princeton: "A young and gifted colleague of mine was recruited on the basis of a pre-print he had written on the quantum theory of measurement, to this day a highly controversial subject... It was all but unheard of for a young physicist to work on something like this. It was not even considered physics. I admired his courage. I do not think that he asked to give a seminar. I think he was told. Oppenheimer was notorious for cutting down seminar speakers who he thought were wasting his time. At this seminar he outdid himself. The speaker had gotten out about five sentences when Oppenheimer said that Niels Bohr had answered all those questions in the 1930s and that the speaker had nothing to add. The seminar came to an end right there. I am glad to report that my colleague published his paper and that he went on to have a distinguished career." The situation only changed starting in the 1960's largely due to the work of John Bell.

Instrumentalism denies all reality to theoretical entities. It does not allow to understand the very success of the scientific explanation of reality, in particular how predictions fit with experiments. If the only coincidence we consider relevant are predictions of the theory for observations, which in the end are data we get through our human senses, we end up subordinating all reality to our sensory organs. That was the position adopted by Hume for whom all knowledge is only about our own mental state. Once all external reference is abandoned, our sensations become the raw materials of the universe. The distinction between sensations that come from external or objective causes and those originating in our imagination becomes blurred. In that sense, Hume himself was forced to resort to the "vivacity" of sensations in order to make a distinction. In this way, physical objects become mere configurations of sensations and their permanence and regularity is essentially circumstantial. The logical development of the subjective idealism of Hume is solipsism, according to which, since all our knowledge is about our own mental states, there is no reason to sustain that anything else exists but the cognizant object, each of ourselves. The problem of understanding these phenomena from a realistic point of view will be one of our key objectives. Let us start by gaining some familiarity with them.

7.1 Non locality and entanglement in quantum systems

The importance of entanglement between the components of a composite quantum system was not immediately recognized. Two papers appeared in 1935 dealing with the phenomenon. In the first one, Einstein, Podolsky and Rosen argued using an entangled system that quantum mechanics was apparently incomplete. Shortly thereafter Erwin Schrödinger published a seminal paper discussing the notion of entanglement. He realized the crucial role of the concept: "[Entanglement] is not one, but rather the characteristic trait of quantum mechanics, the one that enforces its entire departure from classical lines of thought."

Let us attempt to explain this surprising property of composite quantum systems, that is, systems that involve more than one particle. In classical physics any system in motion has perfectly well defined properties for all physical magnitudes. Indeed, given the position and velocity, any magnitude associated with it, energy, momentum or any other, can be computed. In quantum mechanics things are not that way due to the uncertainty principle that does not allow to know the position and velocity of particles with arbitrary precision. On the other hand we have emphasized that quantum systems do not really have physical properties until we measure them. We will say that a given system in a given state has a potential property (for instance "spin up") when, if a measurement of of the magnitude associated with such property were made (for instance a measurement of the z component of the spin with a Stern-Gerlach device), one will observe with certainty as a result such property (the particle would show in the upper branch of a Stern–Gerlach apparatus). In other words, the probability of obtaining that result is one. The quantum counterpart of the classical statement we made at the beginning is the following: *any quantum system with a given state vector always has SOME well defined potential properties.* A system in a proper state, for instance a neutron with spin up, has a well defined potential property. The latter corresponds to the proper result associated with the state. In the example, if we measure the z component of the spin, one will get the result up. Any other property, for instance the other components of the spin, does not have a well defined potential value in that state. For instance we cannot have certainty that the x component of the spin will be up or down. Again, it is the uncertainty principle what ultimately blocks quantum systems from having all classical properties as quantum potential properties defined simultaneously.

Just like any other system, composite ones have some potential properties that are well defined in any given state. However, in entangled systems, it can happen that the composite system has a well defined property while its components have none. In more generality, *an entangled system is one in which the total has more defined potential properties than its parts.*

Before analyzing how can this be possible, it is worthwhile emphasizing that this is something completely new with no analog in classical mechanics. The properties of a classical system of particles can all be known if we know the properties of each individual particle. In quantum mechanics things are different: *the total can have properties that do not result from properties of its constituent parts!*

Let us analyze this in a bit more detail to understand the origin of this phenomenon. We start by discussing how composite systems are described in quantum mechanics. Take two neutral particles with spin. We will distinguish two situations. In the first each particle is in a proper state individually and in the second one they are entangled.

7.1.1 Composite system of two particles in independent proper states

Let us denote as 1 the particle to the left and 2 the particle to the right. We ignore the information about the position of the particles and concentrate on their spins. We will assume that each particle state has been prepared with a Stern–Gerlach device such that particle 1 is in a state "up" along the z direction of spin whereas particle 2 is in a state "up" along the x direction of spin. It is therefore clear that in this case each particle has well defined properties.

To analyze the type of state of this system, we need to dwell a bit more on the nature of quantum states. Up to now we have left things pretty vague, indicating that the states were tools used to compute probabilities. Let us add a bit more nuance to the picture. In its essence a quantum state like that of particle 1 we just discussed $|1, z, \text{down}\rangle$ is a collection of numbers. The numbers indicate the relative probabilities with which the various possible values of the variable in question can occur. In this particular case, two numbers only, referring to the probabilities of spin up or down. In more complicated systems one would have collections of many numbers that we can visualize as a tower, just like one does for the components of vectors. In other systems it can be even more complicated, involving infinite sets of numbers. There are suitable mathematical tools to deal with that point, but we will avoid their discussion in this book, concentrating on simple systems.

The state of a system like the one we started talking about is described by what is mathematically known as a "direct product" of the states. Direct product is an operation that simply puts the states side by side. If one had states that were columns of numbers, one has a new state for the system that is a new column consisting of all possible products of the numbers from one vector with those of the other (so if the original vectors had n components the resulting one will have n^2). States like this can be summed linearly as we discussed before, one simply sums the components of the corresponding towers keeping track of their order (i.e. "which tower is to the right and which to the left"). There are also well defined rules to compute probabilities from such states involving multiple towers of numbers. To simplify the notation, we will not list explicitly the towers of numbers here, but denote things like this:

$$|\psi\rangle = |1, z, \text{up}\rangle |2, x, \text{up}\rangle \qquad (7.1)$$

so on the left we have the composite state $|\psi>$ for the system that is obtained by putting side by side the states for particle 1 and particle 2.

This idea of putting things side by side and keeping track of them like this indicates that in these types of states, both particles operate essentially independently. For instance, if we took the system and ran it through a Stern–Gerlach device that measures the component of the spin of particle 1 along a direction z' forming an angle with z, there is some probability that it will make it through the device with component z' down, and in that case the state after the measurement would be given by

$$|\psi\rangle = |1, z', \text{down}\rangle |2, x, \text{up}\rangle. \qquad (7.2)$$

Notice that nothing has happened to particle 2, its tower of numbers is unchanged by the measurement on particle 1. In this example, the quantum system is behaving like one would normally consider a classical system of non-interacting particles would behave. The properties of the system that are well defined are also properties of its component parts. As in any quantum system one can only make statistical predictions about properties, but in this case the probabilities are computed like one normally does in classical physics. In particular the joint probability of measuring arbitrary properties of each particle is given by the product of the individual probabilities. Just like when one flips a coin there is probability $1/2$ of getting heads, if one flips two and wants two heads, the probability is $1/2 \times 1/2 = 1/4$.

Up to now there is nothing surprising. What is surprising is that what we just discussed is far from the generic situation.

7.1.2 Composite system of two entangled particles

Let us now consider what Schrödinger thought was the most unique aspect of quantum systems: entanglement. It emerges from the fact that states can be linearly superposed as we discussed, including the case in which the state has several towers of numbers.

Suppose we reconsider the system of two particles but now in a state that is a linear superposition of the types of states encountered in the previous section. For instance, a state of the form,

$$|\psi\rangle = \frac{1}{\sqrt{2}}|1, z, \text{up}\rangle|2, z, \text{down}\rangle + \frac{1}{\sqrt{2}}|1, z, \text{down}\rangle|2, z, \text{up}\rangle. \qquad (7.3)$$

Again, these direct product states should be summed keeping in mind the order of the towers, and when they are multiplied by a coefficient like $1/\sqrt{2}$ one multiplies each component of the towers times the coefficient. Recall that when one had superpositions of states, the square of the coefficients in the superposition gave the probability of encountering the system in the state that is multiplied by the coefficient. So in this particular case the system has probability $1/2$ of having spin of the particle 1 up and spin of the particle 2 down and probability $1/2$ of having spin of the particle 1 down and spin of the particle 2 up. That exhausts all probabilities (they sum to one). That means that in particular there is zero probability of finding the system with both spins up. But that also implies that one will never have the certainty that one of the particles has spin up either. If we repeatedly measure spins we will find that half of the time a given particle has spin up and half spin down and the same for the other particle.

Summarizing, a system will have very different behaviors depending on if its components are in independent or entangled states. In the first case each component will have well defined potential properties, and if measured one will obtain results with certainty. If the system is entangled its components lose their individuality and lack well defined potential properties. Only the composite system has them. For instance, in the example, the total angular momentum of the composite particle has a vanishing component along the z direction. This is the case even if the components are widely spatially separated and are not interacting at all. This "holistic" behavior of many quantum systems that are non-separable is not something exceptional but actually it is the most common occurrence in composite systems whose components interact or have interacted. For instance, the electrons in a multi-electron atom are entangled. Or two particles that collide in a particle accelerator may become entangled after the collision. It is a generic property: interactions typically yield entangled states for multi component systems. As we will see, the emergent properties of systems of this kind are key to open the possibility of explaining chemical or biological properties in physical terms.

7.1.3   Non-locality, the EPR experiment and Bell's inequalities

The non-separability we observed is related with another property of entangled systems: their non-locality. Both Einstein and Schrödinger in their 1935 papers questioned what could be the significance of entanglement in our perception of reality. Whereas Schrödinger embraced entanglement as the defining fundamental characteristic of quantum mechanics, and was eager to explore its implications, Einstein basically viewed it as a shortcoming of the theory, exhibiting its incompleteness. He dubbed it "spooky action at a distance". In the end he was proved wrong and today entanglement is at the root of many spectacular applications of quantum mechanics. Einstein's rejection of the action at a distance was quite understandable for someone who had demonstrated the independence of fields from underlying matter supporting them and their role as intermediaries in all interactions, that could not propagate faster than light. We will see however, that the "spooky action at a distance" in the end does not contradict the results of Einstein's theory of relativity.

The 1935 paper of Einstein had Boris Podolsky and Nathan Rosen as co-authors, and is sometimes referred to as the "EPR paper". They conclude that the theory is incomplete and in reality each particle of the system has additional information that is not described in the usual framework of quantum mechanics. They claim that each particle has something like an "instruction manual" about how to respond to each measurement. This would later be referred to as "hidden variables". If we knew the values of those variables the theory would be separable and local, just like classical physics.

John Stewart Bell showed in 1964 that EPR used in their work a notion of locality that was not compatible with quantum mechanics. More precisely, no matter what kind of instruction book is used, the behavior of the systems described by such *hidden variable theories* will not agree with the predictions of quantum mechanics in certain circumstances. In the 1970's Alain Aspect would confirm, in a series of beautiful experiments, that the predictions of quantum mechanics were correct, and therefore ruled out all local hidden variable theories. The failed EPR attempt finally contributed to convince most quantum physicists that the probabilistic character of quantum mechanics is fundamental and is not the consequence of our ignorance of portions of a deterministic theory with additional hidden variables that underlies quantum mechanics.

We will not present EPR's argument in its original form, but use the entangled spins we have been discussing up to now, which allow a simpler illustration of the situation. In this presentation we follow that of Ghirardi.

EPR started from the following assumption, based on the notion of locality considered in relativity: the elements of physical reality of a system cannot be instantaneously influenced at a distance. They consider the entangled state we have discussed and assume that particles 1 and 2 are in two regions A and B very far apart from each other. Let us assume that we use a Stern–Gerlach device to make a measurement in region A. If as a consequence we observe particle 1 having the z component of the spin "up", then according to the rules of quantum mechanics discussed in the previous chapter, there is a "quantum jump" and the state after the measurement will be,

$$|\psi\rangle = |1, z, \text{up}\rangle|2, z, \text{down}\rangle. \tag{7.4}$$

Immediately after the measurement, therefore, the observer located in region A is in position to predict with certainty that the component of the spin of particle 2 in region B will be observed "down". Particle 2 has the potential of having its spin down perfectly well defined and therefore that property should be considered an element of physical reality.

According to the notion of locality we introduced, there is no way of attributing the behavior of particle 2 in region B to the measurement carried out in A since there was no time for the two systems to communicate without something having to propagate faster than light (to emphasize this point is why we assumed the regions were far away). The only thing we can conclude is that particle 2 was predetermined to behave like it did, even before the measurement carried out in A. In the language we used before, particle 2 carried in its "instruction manual" the indication that if a measurement of the z component of the spin took place it should respond "down".

EPR's argument continues like this. If instead of measuring the components along z in A we measured the component along another arbitrary direction z', we would have an up or down answer in A and therefore we would have gathered additional information about the instruction manual of particle 2. Since one does not know a priori which measurement will be made, we conclude that the instruction manual of particle 2 must contain information to respond any question about the spin. In entangled systems, in the interaction process that precedes the entanglement, somehow the instruction manuals become arranged in such a way that the responses of both particles are anti-correlated. When one responds up the other one responds down.

Quantum mechanics does not allow us to know those answers, in particular, the uncertainty principle states that we cannot know simultaneously in a precise way different components of the spin (it is analogous to what we discussed concerning position and momentum not being measurable simultaneously). EPR's conclusion is that quantum mechanics is incomplete since it cannot account for properties that are elements of physical reality that belong to each particle.

It should be noted that ultimately what EPR's argument was trying to salvage was the possibility of thinking that the quantum systems, like the classical ones, have properties that exist even when no one measured them. When we talk about a classical object, we believe its physical properties are always present, even when no one observes them. As we discussed, and although initially it is difficult to accept, in quantum mechanics a particle not observed or not being measured does not have physical properties till the moment in which it is measured. The properties follow from the answers, read in measuring devices, of the micro-systems being measured.

Although Einstein ended up being wrong on this point, his observation was a remarkable contribution to the development of quantum theory (Einstein had made another crucial contribution to quantum mechanics in 1905 when he explained the photoelectric effect through which certain materials generated electricity when illuminated). In this case he started a line of research that shows that one can question the nature of reality, and that contrary to what the positivist logicians and the Copenhagen orthodoxy claimed, such investigations were not "metaphysical" but as Bell showed, could be probed experimentally.

The experiment was based on a result known as Bell's inequality. It stems from a classical analysis of a system of two particles in which one assumes, like Einstein did, that each of the particles have properties that were established even before they are measured. This was actually one in a series of inequalities that are now known generically as Bell's inequalities that are satisfied by systems with local reality like the one proposed by EPR. Again, we will not discuss the original version, but one that is simpler to explain, based on an inequality proposed by Mermin (1985). As has become customary in this field, we will assume we have two experimentalists named Alice and Bob. A third experimentalist (Kim) prepares a pair of neutrons in an entangled state like (7.3), one with spin up and the other down along a random direction, and sends one to Alice and one to Bob. The procedure is repeated a large number of times. Alice has a Stern–Gerlach device that she can rotate around its center, and can set it to three different directions forming angles of 120 degrees with each other, as shown in the figure. Let us call these three positions 1, 2, and 3. In each of the positions, the neutron sent will either make it through the Stern–Gerlach device up or down. So the instrument has a three way switch that allows to choose the direction and a little light that is normally green and turns red when a neutron makes it through the Stern–Gerlach device up and green when it is down. Bob has a similar device, but the light colors are chosen in an opposite way, that is, it will turn green when up and red when down. When Alice receives the particle she decides at which position to set the polarizer (1, 2 or 3) randomly, say by spinning a three sided teetotum. Bob proceeds in a similar way with his instrument. They write down their results. After recording many particles, the experiment ends and they compare notes. They realize that events in which the lights are of the same color, either red or green, occur with the same probability.

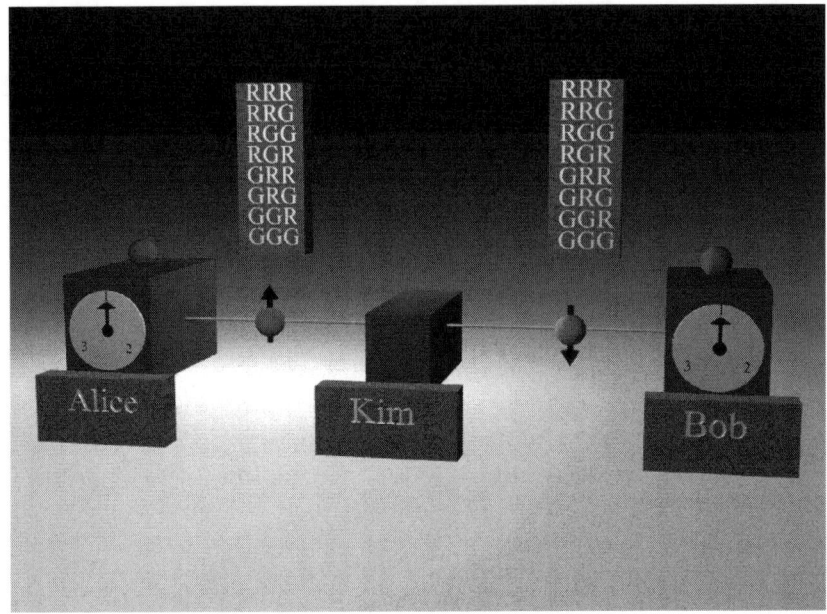

Figure 7.1: The Einstein–Podolsky–Rosen (EPR) experiment that leads to Bell's inequalities. Shown are the flying neutrons and on top of each the "book of answers" that each should carry in a hidden variable theory.

Quantum mechanics accounts for this fact precisely[1]

Let us now try to account for the experiment assuming each neutron

---

[1]In detail: suppose both chose the same position (either A, B, or C); then if one light is red the other will also be red given how the state was prepared and the convention we chose for the lights. This will happen only 1/3 of the times, since there are three possibilities for Bob for a give choice of Alice. If they chose different positions, it means both detectors are at 120 degrees with respect to each other. Say Alice chose A. Then Bob chose B or C. The probability of agreeing then is given by $\cos^2(120°/) = 1/4$ (just like in polarizers and light, the probability of making it through a Stern–Gerlach device at an angle is proportional to the cosine of the angle squared, there is a factor 1/2 in the angle for neutrons because they are spin 1/2 particles) for each B or C. Bob chooses 2/3rds of the time an orientation different from that of Alice. Every time he does, the probability is 1/4 as we saw. Therefore the total probability is $1/3 + (2/3) \times (1/4) = 1/3 + 1/6 = 1/2$. This is the value observed experimentally.

carries a book with preselected answers, what is commonly known as a hidden variable theory. So every time Kim prepares a pair of particles, she has to tell them through these hidden variables what result to give if either Alice or Bob choose 1, 2, 3. In other terms, Kim implicitly "writes the book" when preparing the state. So there are eight possibilities for those variables (RRR,RRG,RGR,GRR,GGR,GRG,RGG,GGG). Kim prepares the particles with equal probabilities for all these choices and sends particles to Alice an Bob with the same value for each choice, such that if they choose to measure the same property, they will get the same result. Consider a particular instruction set, say RRG. Should both particles be assigned same set, they will flash the same color if the polarizers are set in one of these combinations: 11,22,33,12 or 21. Evidently similar results hold for sets GRR,GGR,GRG,RGG. They will flash the opposite color for 13,23,31,32. So of the nine possibilities they flash the same colors 5/9-th of the times. In addition to this they would flash the same color all the time if the sets were GGG or RRR. So a theory of this kind predicts that *at least* 5/9-ths of the time the lights will flash the same colors. But an experiment similar to those of Alain Aspect would have shown that the correct result is the one predicted by quantum mechanics: they do so 1/2 of the time.

Bell's inequalities are relations between probabilities in systems that have to be satisfied if the system obeys locality in the sense of Einstein (i.e. it has hidden variables) and that quantum mechanics and experiments violate. They include the relations about the probabilities in the experiment we discussed (i.e. the probability being at least 5/9ths). Bell's conclusion is that quantum mechanics cannot be reconciled with the hypothesis that the particles carry local elements of reality with them before they are measured. This result shows that the view of reality that quantum mechanics will lead to must differ radically from that one held by physicists and others before its discovery. The experimental confirmation of Bell's result indicates that an urgent rethinking of our view of reality is needed.

We now know that the problem laid out by Einstein was not metaphysical in nature and admitted a definitive unambiguous answer by quantum theory. *Einstein in fact was wrong: quantum particles do not have properties before they are observed.* Abraham Pais begins his book "Subtle is the Lord" by saying "I recall that during one walk Einstein suddenly stopped, turned to me and asked whether I really believed that the Moon exists only if I look at it." If Einstein's question had referred to micro-systems, we today know the answer and it is that definitely electrons do not have a spin until it is measured. For the Moon, as Ghirardi observes, the problem is more subtle because "The consideration of such a system requires a preliminary solution of the problem of the quantum description of macroscopic systems." This problem is usually known as the measurement problem and is perhaps the central mystery of the usual presentations of quantum mechanics. In the second part of this book we will discuss several realistic interpretations of quantum mechanics that attempt to solve this problem and we will see that the answer to Einstein's questions is that, indeed, like most macroscopic objects, the Moon exists when no one sees it.

Quantum non-locality has at least three properties that also distinguish it from any classical behavior, for instance, that of Newtonian gravity and its force law:

- *Its effects only involve the entangled particles.* If we consider a pair of entangled particles in a group of many particles, only this pair gets affected when we measure one of its particles.

- *Its effects are distance-independent.* If the neutron of the entangled pair we considered in this chapter in state $|\psi\rangle$ is measured with z component up, the other neutron, irrespective of how far it is, will be measured in the z down state, since the state after the measurement is $|1, z, \text{up}\rangle |2, z, \text{down}\rangle$.

- *There is no way of turning this into a mechanism for transmitting information, so no violation of relativity takes place.* The speed of light is still the maximum speed at which messages can be transmitted. Although Alice or Bob can choose what to measure, they cannot control the result of the measurement. Only a posteriori if they get together and compare notes they can realize that when they decided to measure the same property their results were correlated.

Asher Peres (1984) has elaborated a dialog about the implications of Bell's inequality following the model of Galileo's *Discorsi*. In it, Simplicius believes in EPR's hypothesis and considers quantum mechanics is incorrect. Salviati basically believes that quantum mechanics is incomplete, whereas Sagredo understands the profound implications of the required conceptual shift. We highlight the main points of the dialog.

Referring to the EPR argument and the inequality Bell obtains from his hypothesis it says:

"Salviati. ...there must be a flaw in this argument. Either it is wrong that the observers have a free choice among the alternative experiments [A,B,C] (namely, for each pair of particles, only one of the four experimental setups is compatible with the laws of physics, —the others are not for reasons unknown to us) or it is wrong that each photon [or other spinning particle] can be observed without disturbing the other photon. Take your choice.

Simplicius. Both alternatives are distasteful. I prefer classical physics.

Salviati. I again insist: This difficulty is not the fault of quantum theory. Only experimental facts are involved here. So indeed we have a paradox.

Sagredo. There is a paradox only because you force on this physical system a description with two separate photons (or electrons). These photons exist only in your imagination. The only thing you have really prepared is a pair of photons, in a ...[definite total] spin state. The pair is a single, indivisible, nonlocal object..."

7.1.4 The meaning of quantum states: no cloning and teleportation

We saw that given a system about which we have no information, carrying out measurements on its state does not allow to know the state prior to the measurement. Let us recall that the rules of quantum mechanics say that if a photon goes through a polarizer aligned with the x axis, it emerges in a state $|e_x\rangle$. It does not matter what state it was in prior to the measurement, it undergoes a sudden change and ends up in $|e_x\rangle$. Therefore there is no way with a single measurement to determine the state of the photon prior to it, since the measurement destroys the state and does not give enough information to determine it. Only if we had a very large set of photons in the same state and measured all their polarization properties we could, through a statistical analysis of the results, determine the state in which those photons were. Usually physicists work with the reduction postulate (what we called the rule v) in the previous chapter) to prepare the photons in a known state, e.g. $|e_y\rangle$ and then study its future behavior.

This limitation in the information we can acquire about a quantum system is related to what is known as the *no cloning theorem*. We will not offer a proof of it, but let us discuss its content. Let us assume that initially a photon is in an unknown state of polarization $|\psi\rangle$. Is it possible to put a second photon in the same state, so in the end we have two photons in the same state? Let us call it $|1, \psi\rangle|2, \psi\rangle$. Could a measurement or using the Schrödinger equation somehow allow us to clone a state? The answer is negative, one cannot clone quantum states about which we have no information, that is, we do not know how they were prepared. This is not only true for photons but for any quantum system. The implication of this is important. If one could clone quantum states, one could, starting from an unknown state prepare a very large number of copies, carry out measurements and learn as much as one wants about the state. Cloning would allow to determine a state without perturbing it.

One can indeed clone a state, provided one destroys the original state. This process is known as *teleportation* and was discovered in 1993 by Bennett and demonstrated experimentally in 1997 by Zeilinger and De Martini. It is remarkable that although one cannot know the information of a state through a measurement, one can teleport the state with a single measurement and some operations. It should be noted that a state characterizes the disposition of a system to give a set of results. What will be teleported are those dispositions, not the system itself. If we consider for instance an hydrogen atom in certain state, one can put another system composed by one electron and one proton in the same state at the cost of perturbing the state of the original atom. To quote Asher Peres, when asked by a reporter if quantum teleportation could teleport the soul as well as the body, Peres answered, characteristically, "No, not the body, just the soul." We are not going to explain how teleportation works. We will only mention that in order for Alice to teleport a state to Bob, both must share one of the components of an entangled system and that the quantum information about the state is transmitted using the entangled pair. Alice must transmit to Bob using classical signals like those of a radio or a telephone information about certain results of measurements that she had to carry out.

Teleportation therefore does not involve the transfer of matter, only of the disposition of a quantum state. If one were talking about atoms, for instance, what would happen is that the state of Alice's atom would be transferred to Bob's atom, not the atom itself.

With this we conclude a first overview of the surprising properties of entangled systems. The holistic character of them will play a central role in the analysis of the emergence of complex phenomena like the chemical or biological ones we will study later on. This is the new element that is a game changer that allows to understand the emergence of complex systems, all with causal capabilities on their constituents. This phenomenon, known as downward causation, can play —as we shall see– a key role in the reformulation of old problems like those of the effects of the mental over the physical.

7.2   The problem of quantum measurements

You may have noticed that the rules of quantum mechanics establish two apparently incompatible types of evolution. While they are not being measured, the quantum states evolve deterministically obeying a differential equation known as Schrödinger's equation. For some reason in measurement apparata this description breaks down and must be replaced by another one in which states change abruptly. They transition from an initial to a final state associated with the proper result of the measurement, that is, the event that is observed in the measurement device. These two evolutions are strictly incompatible, one can show that the type of evolution that occurs in a measurement cannot be described with a Schrödinger equation.

John Stewart Bell (1989) describes the problem in the following terms. "There is a fundamental ambiguity in quantum mechanics, in that nobody knows exactly what it says about a particular situation, for nobody knows exactly where is the boundary between the wavy quantum world [described by states that evolve with the Schrödinger equation] and the world of particular events [like the ones that occur in measurement devises] is located. That for me is the problem of quantum mechanics. It is not a problem in practice —in practice, we always can take this boundary judiciously so that moving it a bit one way or other doesn't matter. But every time we put that boundary in —we must put it somewhere— we are dividing the world arbitrarily into two pieces, using two quite different descriptions, one on one side one on the other."

All this becomes problematic when we attempt to consider quantum mechanics as a description of reality beyond that of the instrumentalist viewpoint. When we want to think about what kind of world quantum mechanics talks about and not only account for experiments made by physicists. The problem is not just that one has two different forms of evolution. The key difficulty, which we have discussed in some detail and worried Einstein, is in the fact that there do not exist objective elements of reality until a measurement is carried out and events are produced in the measuring device.

The problem originated with the very birth of quantum mechanics almost ninety years ago and has seen several proposals arise for its solution, up to now all unsatisfactory for one reason or another. In a nutshell, the issue is that quantum mechanics allows to predict the behavior of a physical system subject to measurements. However, in its current formulation, it does not allow to say what happens when an observer is not involved, that is, when there are no measurements on the system. In spite of the fact that a good fraction of 20th century technological advance is due to quantum mechanics and that the theory has predictive power that sets the basis for chemistry and molecular biology, it has not been possible to reach a consensus about its interpretation. The ultimate goal would be to elaborate a coherent vision of reality valid both for microscopic systems which we probe through specialized measuring devices and macroscopic systems accessible to our everyday experience.

The difficulty in establishing a boundary that separates when we have to abandon the deterministic and linear description given by the Schrödinger equation, where there are no objective elements of reality, and transition to the description in terms of events and quantum jumps that occur in measuring devices is illustrated by a thought experiment known as *Schrödinger's cat*.

Schrödinger himself presents the problem as follows: "One can even set up quite ridiculous cases. A cat is penned up in a steel chamber, along with the following device (which must be secured against direct interference by the cat): in a Geiger counter, there is a tiny bit of radioactive substance, so small that perhaps in the course of the hour, one of the atoms decays, but also, with equal probability, perhaps none; if it happens, the counter tube discharges, and through a relay releases a hammer that shatters a small flask of hydrocyanic acid. If one has left this entire system to itself for an hour, one would say that the cat still lives if meanwhile no atom has decayed. The psi-function of the entire system would express this by having in it the living and dead cat (pardon the expression) mixed or smeared out in equal parts. It is typical of these cases that an indeterminacy originally restricted to the atomic domain becomes transformed into macroscopic indeterminacy, which can then be resolved by direct observation. That prevents us from so naively accepting as valid a 'blurred model' for representing reality. In itself, it would not embody anything unclear or contradictory. There is a difference between a shaky or out-of-focus photograph and a snapshot of clouds and fog banks."

Schrödinger in fact questions the orthodox instrumentalist view repeatedly professed by Bohr, arguing that although our intuition says no observer can be in a mixture of states, the cat can be in such a mixture. The question of whether the cat is or isn't a measuring device is equivalent to the question about when we are supposed to apply the Schrödinger evolution and when the reduction postulate.

Let us lay out the problem in slightly more technical terms. We represent the state of the radioactive atoms when some time has elapsed as $(|E\rangle + |NE\rangle)$ where $|E\rangle$ represents the state when the radioactive atoms have emitted an alpha particle and $|NE\rangle$ the state when no particle has been emitted. The linearity of Schrödinger's equations applied to the complete system including the cat leads to a state $(|E\rangle|\text{dead}\rangle + |NE\rangle|\text{alive}\rangle)$. If one were to consider the cat a measuring device it would either be in the state $|NE\rangle|\text{alive}\rangle$ or in $|E\rangle|\text{dead}\rangle$, but it cannot be in a superposition. The theory does not specify when one description ends and the other one starts. Summarizing, the reduction postulate contradicts Schrödinger's equation, which describes the evolution of quantum systems. The theory cannot, without some type of additional hypothesis, describe how a definite outcome emerges from the interaction of a quantum system with a measuring device.

To establish in an ad hoc way that the theory only is valid for micro-systems because we cannot make sense of having a cat potentially dead and alive at the same time is not too satisfactory. Especially today, when we have many macroscopic systems like superfluids or superconductors where quantum superpositions take place.

We are therefore facing the problem that Ghirardi calls the *macro-objectification* of properties, characterized by the passage from a world of potentialities to one of realities: "how, when, and under what conditions do definite macroscopic properties emerge (in accordance with our daily experience) for systems that, when all is said and done, we have no good reasons for thinking are fundamentally different from the micro-systems of which they are composed."

This is the great mystery that any realist physicist or philosopher has to face and to whose solution we will devote more time in the second part of this book.

Chapter 8

Towards a complete unification of the conceptual frameworks, including all forms of matter

8.1 Introduction

In general relativity, the most developed form of classical physics, there exist three forms of matter: particles, fields and space-time itself, which is dynamical and can convey energy. With quantum physics, this proliferation of the forms of matter begins to revert itself. Indeed, the wave-particle duality is a hint that particles and fields are two manifestations of the same object. As we will see, this initial observation is confirmed by a new development in physics that took place throughout the 20th century: the quantum theory of fields.

Until the 1920's the only known particles were the electrons, discovered in 1895 and the photons theoretically introduced by Einstein in 1905. Ernest Rutherford proposed the existence of the protons and neutrons in 1920 in order to try to explain atomic nuclei. Neutrons

were only observed experimentally in the 1930's. In the early 1960's
it was proposed that the proton and neutron are not elementary particles, but are composed by quarks, and this was experimentally verified
in the late 60's. Electrons, photons and quarks are particles that are
truly fundamental, that is, they are not believed to be composed by
other particles, at least in the paradigm known as the Standard Model
of Particle Physics of today[1]. They can be thought of as point particles
although we have learned, since they are quantum objects, that they
behave dually: sometimes as particles, sometimes as waves. We will
discuss the duality later on.

Photons, as we discussed, are particles associated with electromagnetic radiation. Particles associated with gravitational waves are known
as gravitons and particles associated with the nuclear forces are known
as gluons. Photons, gravitons and gluons are generically known as *vector
bosons*[2].

Quarks have never been observed in isolation, they only arise in nature composing other particles like the proton and neutron. This is explained by a process known as *confinement*. Unlike the electromagnetic
force, that decreases with the distance squared, the strong interaction
among quarks, which in turn is responsible for the nuclear forces among
protons and neutrons, increases with distance. That prevents quarks
from flying away, no matter what state of motion one can set them into.
In spite of the fact that we cannot see quarks in isolation, since the late
60's we have indirect experimental evidence for their existence and how
they are kept together by strong interactions mediated by the vector
bosons known as gluons.

The unification started by quantum mechanics was incomplete. The
final understanding of the quantum behavior of the electromagnetic,
weak and strong interactions required a treatment that is both quantum mechanical and special relativistic. It is expected that a quantum
theory of gravity will also require a treatment that is quantum and
(general) relativistic. One can quickly understand that electromagnetic
fields, that propagate at the speed of light, require relativity for a proper
quantum description. Electrons can be accelerated to high speeds, where
one cannot ignore relativistic effects, and it was noted that particles are

---

[1] There exist proposals of unified theories where some of these particles become
composite, but they are not widely accepted yet.

[2] Vector in this context refers to the fact that they mediate the interactions, the
gluons and photons are spin one particles, the graviton has spin two.

created or destroyed in the process. High energy particle physics explicitly shows the equivalence of mass and energy: although energy is preserved in high energy processes, the number of particles can change as long as part of the energy morphs into the rest energy of the created particles. For a particle of mass M it requires having at hand at least Mc^2 of energy to create it. As we mentioned, the final unification of the three types of matter previously mentioned, will require including general relativity. It will allow to complete the unification of particles, fields and space-time as different manifestations of the same entity. Although we have not achieved a fully unified theory up to present, current approaches being considered are all quantum in nature.

8.2 Quantizing electromagnetism

Quantum mechanics, that by 1927 had been developed by Bohr, Heisenberg, Schrödinger and a handful of other physicists, describes systems that in certain limits reproduce the behavior of non-relativistic particles. It should not surprise that one of the first difficulties found with the new theory was when a complete quantum mechanical treatment of electromagnetism was attempted. This required a relativistic treatment. The first step was to try to quantize the free electromagnetic field, that is, without charges present. As in any other quantum system, the states of the electromagnetic field characterize its disposition to exhibit certain properties in measurement processes. Among them are the ones associated to the classical magnitudes that characterize the electromagnetic field, like the values of the electric or magnetic fields. But they also include others associated to the number of particles that can be detected by a photodetector. They are particles with zero mass but with non-vanishing energy. Their angular momentum is analogous to the spin we discussed before, and its value is one in units of $h/(2\pi)$. This spin is associated, from the point of view of classical physics, with the polarization of the electromagnetic waves. These particles are known as photons. As a consequence, treating quantum mechanically the electromagnetic field allows to understand the wave/particle duality as just two different behaviors of quantum electromagnetic fields. Sometimes they may behave like a wave, sometimes like a particle.

The corpuscular and wave-like behaviors of light are complementary and satisfy a relation akin to the uncertainty relation between position and momentum for an electron. If one prepares the state of a field such that it has an electric or magnetic field defined with great precision, if one attempts to measure the number of photons involved, one will observer large dispersions. Conversely, if one prepares a state with a fixed number of photons and measures the electric or magnetic field, repeating the experiment on the same state many times will yield values very different of the fields for each experiment. These types of states, with a well defined number of photons, behave very differently than ordinary electromagnetic waves.

8.2.1   Recovering the classical behavior of fields: coherent states

Neither the states with perfectly well defined fields nor the states with well defined number of photons are good approximations to the classical behavior of the fields. The states with totally well defined electric field have large uncertainties in the values of the magnetic field and the energy. States with a well defined number of photons have a well defined energy but large fluctuations in the magnetic and electric fields. This leads to question of which are the states that we observe in our macroscopic experience when we deal with electromagnetic fields. To attempt to answer this, it is good to focus on the states that better describe classical free particles. It is clear that they cannot be states with well defined positions because that would require a large uncertainty in the velocities and that is not what we observe in classical free particles. The idea to approximate the classical behavior is to admit a certain degree of uncertainty both in position and in momentum and to try to get the smallest amount of uncertainty compatible with Heisenberg's uncertainty principle. Sometimes people talk of recovering the classical behavior when one accepts a degree of *coarse graining* of the measurements, i.e. accepts a certain degree of uncertainty in them. Given that Planck's constant is very small, the uncertainties of macroscopic objects can be much smaller than the macroscopic scales and the quantum system will have a behavior that departs very little from the classical one, at least for a limited time. So little that we do not notice it. States that achieve this are called minimum uncertainty states.

The states that better approximate the behavior of a classical electromagnetic wave are minimum uncertainty states. If the light is sufficiently intense, its behavior is similar to that of an electromagnetic wave with electric and magnetic fields oscillating in time. Roy Glauber was the first to give a complete description of these states, which are known also as *coherent states* because all photons move in phase together to produce the classical behavior. Lasers provide the closest physical implementation of coherent states. The light emitted by a lightbulb differs significantly from Glauber's coherent states. It is composed by photons that have different frequencies and polarization in a process that is random in phase and time. This is easy to understand given that a lightbulb emits light by heating a filament to high temperature, a state in which its molecules vibrate violently which leads them to emit photons. Each photon is emitted from a different molecule and this leads them to have different properties. In coherent states the uncertainty principle leads to the energy, and therefore the number of photons, to have large uncertainties.

8.2.2 Quantum field theory

In the chapter on the origins of quantum mechanics we saw how the quantum behavior of individual photons that traveled through polarizers reproduced, when one considered a large number of photons, the predictions of classical electromagnetism. In the two slit experiment we saw how photons distribute themselves on the photographic plate reproducing the intensity curve predicted by interference that follows from Maxwell's equations of classical electromagnetism. If the two slit experiment is repeated with electrons the wave/particle duality still exists and the curve of intensities is proportional to the wave function squared. The evolution of the latter is given by Schrödinger's equation. If the electrons move at speeds closer to that of light, then one will have to use Dirac's equation instead.

A first conclusion that may appear rather surprising is that one could consider Maxwell's equations to be the quantum equations that describe the wave-like behavior of the photon. The renowned particle physicist Steven Weinberg observes about this that the "photon is the only particle that was known as a field before it was detected as a particle". The quantization of the electromagnetic field allows to treat the processes of creation and annihilation of photons and, in general, systems with a large number of photons. If we wish to treat processes in which the number of electrons can change, the road seems to be to proceed analogously and to quantize the Dirac spinor fields.

Quantum field theory methods could therefore be applied to the equations that arise from treating quantum mechanically particles like the electron. Dirac called this procedure "second quantization" because it consists in treating as classical equations that were already quantum mechanical and quantize them again. These set of techniques is known today as quantum field theory (QFT).

The states of a QFT will characterize the disposition of the system to produce a given number of particles in a detector with certain energies and momenta. In general we will have, as is typical in quantum mechanics, some uncertainty in the number of particles and their properties. The wave-like or particle behaviors are now merely different responses that a quantum field can give in a measurement process. One can therefore claim that with QFT we have a unique conceptual framework to understand particles and fields. Both manifest themselves as events in detectors, like in a Geiger counter or in the traces that a particle leaves in a bubble chamber.

The events involved in the detection are always the product of an interaction with measuring systems that are macroscopic. In the case of the bubble chamber, the drops of liquid are vastly larger than the particles that leave the trace and behave macroscopically. As we shall see many interpretations of quantum mechanics provide criteria for the occurrence of events in macroscopic systems. Both quantum mechanics and quantum field theory are formulated in terms of the same concepts, systems, states and events, and present the same interpretational issues concerning the criteria for the occurrence of events.

8.2.3   The vacuum state and virtual particles

In QFT the definition of particles depends on identifying a state called the vacuum that has the minimum possible energy. A unique determination of such a state is possible when gravitational fields are so weak they can be ignored. In that case the vacuum state is the same for all inertial frames. The vacuum is the state in which all detectors in uniform motion (without acceleration) do not detect any particle.

The vacuum is far from an empty region. According to quantum mechanics we will see that in reality at every point of space, including those in which there are no detectors, there is an intense activity where particles are created and annihilated constantly. Before going into detail, the reader should not be surprised that in every region of

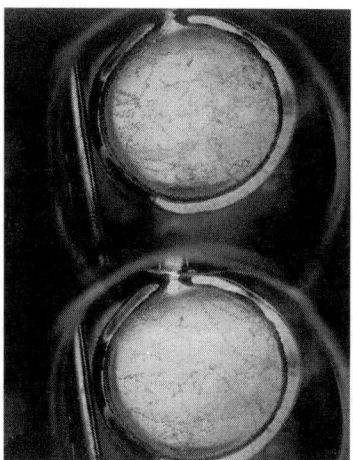

Figure 8.1: Traces left by particles in a bubble chamber. The latter is a device filled with superheated liquid hydrogen. A superheated liquid is in a state that is beyond its boiling point but has yet failed to boil. Small perturbations can lead it to boil. When a particle shoots through it, it provides the small perturbation and traces of vapor are left along the trajectory. The picture shows John Wood's 1.5 inch chamber in 1954.

space there exist physical processes. In fact, with general relativity we had already noted that space-time itself has dynamical character and is affected by matter and in turn influences it. The vacuum of QFT is another realization of how erroneous is to assign to the vacuum the properties of a mere geometric space. *Just like space-time is a physical entity, so is the vacuum. We live in a deep sea of matter that is in perpetual motion.*

Behind this permanent microscopic activity of space-time is Heisenberg's uncertainty principle. Even in the vacuum the uncertainty relation between time and energy tells us that for short intervals of time the energy of a system can suffer significant fluctuations. A large enough fluctuation can permit the creation of a pair of particles. One would need a fluctuation larger than twice the rest energy Mc^2 of the particles created. These types of particles that sprout from the vacuum spontaneously are known as virtual particles, and do not live long, in order to comply with the uncertainty principle. The shorter the time, the larger the energy fluctuations. A new phenomenon occurs when we consider fluctuations that live shorter than a *Planck time*, which is equal to 10^{-44} seconds (that is zero followed by a decimal point followed by 43 zeros and one). In that case, the fluctuations are so large they can produce black holes and disrupt the continuity of space-time. This frenzy of exchange of energy gets more intense the smaller the scales one probes. When space-time itself starts to fluctuate and loses its continuity one talks of a *space-time foam.*

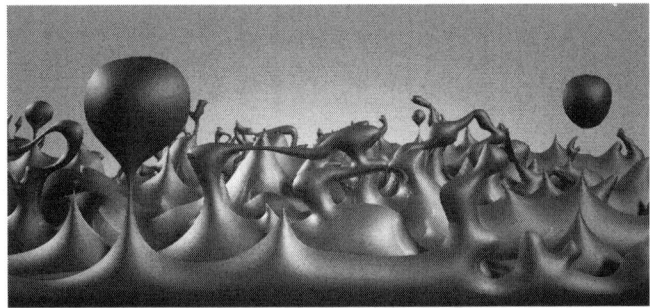

Figure 8.2: As one probes space-time at shorter and shorter lengths, it appears less and less smooth. Eventually at Planck length (10^{-33}cm) space-time appears like a sea with space-time foam, as John Archibald Wheeler called it. (Credit: NASA/CXC/M.Weiss)

Planck's time is a quantity determined by the three most important universal constants of physics, the speed of light c, the constant of gravitation G and Planck's constant h. Each of them defines the scale in which certain phenomena become important. When speeds approach c one needs Einstein's theory of relativity for a correct description. When gravitational forces controlled by G become important we need general relativity. When the scales of a system are such that the product of the accuracy of measurement of time times the accuracy of measurement of the energy becomes small enough to be comparable to Planck's constant h, quantum effects are important. Analogously, Planck's time Planck's time (given by $t_P = \sqrt{\hbar G/c^5} \sim 5.39 \times 10^{-44}$s) characterizes the time scale in which space-time loses continuity. As in the other cases, this constant characterizes when a new theory is required. To completely understand physics at such scales one would need a quantum theory of gravity. If one wishes to talk about the spatial scale at which space-time loses continuity one has Planck's length $\ell_P = c\,t_P = 1.6 \times 10^{-33}$cm.

It should be noted that in QFT the classical notions of particles and fields appear as behaviors of the states of the quantum fields that manifest themselves in the events they produce in other systems. For instance, the charged particles in their motions produce bubbles in a bubble chamber that leave a trace as the one we showed in the figure. *The fundamental objects of contemporary physics are not the particles but the events produced by the states.* Particles have a relational character. Physicists have noted the relational nature of the world long ago. The very existence of an object arises always in relation to another object. Whenever an object is seen it is in interaction with others, nothing can be said to be intrinsic to an object. Suppose one has a detector placed in a vacuum and it does not detect anything. One could conclude that there are no particles in that place. But Bill Unruh demonstrated in the 70's that if one places a detector accelerated with respect to the previous one, it will see particles. The existence of the particle depends on the state of acceleration of the detector. Should one say that there is or that there is not a particle in that region? The way to think of this is to consider a system that pervades all of space (the field) that interacts with another system (the detector). The answer given by the system to two different detectors can be different. The result (the existence of the particle) clearly depends on the interaction, it is a relational concept.

As we saw, the vacuum is such that particle detectors in inertial systems do not see particles. The fact that accelerated detectors register them suggests that the notion of particle in QFT defined with states in a space-time without gravitational forces should be modified if one studies QFT in a curved space-time. We will not study this in detail. We will just mention that time-dependent gravitational fields produce particles. They are also produced close to the horizon of a black hole, in the so-called Hawking effect. The particles produced carry energy away from the black holes. The latter evaporate and in the end disappear. For astrophysical sized black holes, the process is extremely slow and unobservable by many orders of magnitude.

8.2.4   Interacting fields and infinities

Physically relevant QFTs describe particles that interact with each other.
For instance, electrons and positrons interact with photons. The QFT
that describes that interaction is known as quantum electrodynamics
(QED). Dealing with interacting QFTs has been problematic to accom-
plish in closed form, due to the complexities involved. Approximation
techniques have to be used. The main one is known as perturbation theory. The coupling of electrons and photons is governed by the electric
charge. More precisely the constant that appears when the equations
are laid out is $\alpha = e^2/(hc)$. Such a constant α is dimensionless and approximately takes the value of $1/137$ and is known as the *fine structure
constant*. As can be seen, it is a small number. Therefore one expects
that the coupling of photons and electrons and positrons will not modify
very much the individual theories one would have if they were not interacting. The idea is to build the interacting theory by starting with the
non interacting theories and incorporating the interactions gradually.
One starts by considering terms that are linear in α and then consider
terms that go as α^2 which will be 100 times smaller, etc. The risk is
that the method is not guaranteed to work. We only know that the
second batch of terms is proportional to α^2. That does not guarantee
that they are smaller, the proportionality constant could turn out to be
large. In spite of these caveats, the method does work, at least for QED,
and produces results that have been verified experimentally with enormous precision. The detailed technique for doing perturbation theory
is known as Feynman diagrammatics. The calculation can be carried
out to whatever level of accuracy needed by including terms that are
proportional to higher powers of α but the calculations quickly become
prohibitive. Rarely one needs calculations beyond α^4 to match current
experiments.

Although the current status of the method is as described, when it
was first tried out in the 1930's it faced formidable obstacles. In fact it is
kind of remarkable that in the end it worked at all. The main difficulty
encountered is that the coefficients of the various powers of α, far from
being small, actually diverged in many cases. The problem is related
to a property known as renormalization. The idea of perturbation is
that the fields are well described to a first approximation by the non-
interacting equations. For instance for the electrons, it would be the
Dirac equation with the mass that enters in it given by the mass that

we measure experimentally for the electron. However, since the electrons interact with virtual particles produced in the vacuum via their own electromagnetic fields, the naive notion of free particle that one uses to get perturbation theory started (the one described by the Dirac equation alone) is woefully incorrect since it ignores those interactions. The particles we start the theory with ignore interactions with the fields, the particles we see even in isolation (and which we may use to measure the mass of the electron) are constantly interacting with the fields. The interaction of a charge with its own field actually increases as one gets closer to the charge and actually diverges when the distance to the charge goes to zero. It is unsurprising that such object would therefore involve infinite corrections with respect with the non-interacting object described by the Dirac equation alone. In fact, the charge and mass of isolated electrons differ by an infinite amount from those described by the non-interacting Dirac equation. The bottom line is: what appeared as a free particle in starting perturbation theory is not what we consider a free particle in the lab and on which we can make measurements of mass and charge.

The phenomenon of the *renormalization* of fundamental constants was developed into a full procedure for treating QFTs in the 1930's and 40's by Enrst Stückelberg and later by Julian Schwinger, Richard Feynman, Shin'ichiro Tomonaga and Freeman Dyson. Its central idea is to start the perturbation theory with values for the charge, mass and others that are not the ones physically observable. At the initial moment they are just free parameters. One then adjusts those parameters to eliminate the infinities from the interacting theory. This may sound very counterintuitive. There are infinite terms in the perturbative theory and only a finite number of parameters to adjust. Remarkably, for several theories the procedure works. One can eliminate all infinities adjusting a handful of parameters. QED is an example of such theories.

As can be inferred from the previous statements, it is not easy to find theories that are renormalizable, it is a challenge to have all infinities be swept away by redefining a few parameters. An example of a non-renormalizable theory is general relativity, at least in our current understanding. In that case, there are new types of infinity at each stage and one quickly runs out of constants to reabsorb them.

The fact that one can make sense of renormalizable QFTs and they can be used to make great verifiable predictions for experiments does not imply that they are free of difficulties. The infinities that appear

are symptoms that the idealized treatment in which one assumes that
particles are material points and space-time is continuous at infinitely
fine scales is problematic. Dirac says "The rules of renormalization
give surprisingly, excessively good agreement with experiments. Most
physicists say that these working rules are therefore, correct. I feel that
it is not an adequate reason. Just because the results happen to be in
agreement with observation does not prove that one's theory is correct."
Renormalization can therefore be viewed as a practical way of computing physical magnitudes of relevance without having to worry about
what happens at very small scales. In order to understand the situation
better one would need a better handle on the microscopic structure of
space-time and explain why in renormalizable theories the microscopic
effects can be ignored. Any treatment of the ultramicroscopic behavior of space-time requires both a relativistic and quantum treatment of
gravity.

8.3 The issue of unifying quantum mechanics and gravity

8.3.1 Towards a complete physical description of the world: quantum gravity, loops and strings

The conceptual revolution started by quantum mechanics and relativity
is unfinished. Only two of the three fundamental material objects: particles, fields and space-time, have been quantized. QFT has achieved
that goal for fields and particles. For space-time we currently only have
the classical outlook provided by general relativity. Therefore, a unified
treatment of all fundamental forms of matter has not been achieved. If
one approaches the problem from the fundamental interactions of nature, QFT has unified three of them (the electromagnetic, weak and
strong interactions) into what is known as the Standard Model of particle physics. But gravity has yet to be incorporated into the framework.

In addition to this, the two main paradigms of physics, general relativity and QFT have internal issues. In general relativity strong mathematical theorems proved in the 1960's and 70's predict that space-times
become singular under rather general conditions. The Big Bang we believe to be present at the origin of the universe and the singularities
that arise inside black holes are examples of such singularities. In gen-

eral terms, a singularity is associated with a divergence in quantities that
indicates the theory has been taken beyond its realm of applicability.
Close to such singularities one usually encounters energy densities that
are not compatible with a completely classical treatment. It is therefore
reasonable to expect that a theory that unifies QFT with general relativity could offer new insights, and perhaps eliminate the singularities
altogether. Similarly, the quantum theory of fields has the problem that
many fundamental quantities take infinite values as we have already
discussed. To address that problem one needs to identify the quantum
structure of space-time at ultramicroscopic scales and that would also
require a quantum treatment of gravitation.

The absence of a completely satisfactory theory of quantum gravity,
needed to place all physical phenomena and matter on the same footing,
leaves us facing two theories with mathematical structures and physical
concepts that are very different. General relativity and QFT have had
such an empirical success that at present we know of no single physical
phenomenon that cannot be explained by them. We know, however,
that one must carefully limit the regimes in which these theories are
used. It might be considered that to find a theory that unifies both is
an impossible task until we have experimental evidence of phenomena
that will require to push both theories out of their current realm of applicability. It appears that the current situation is a theorists' dream:
with no experimental data to constrain them, it should be easy to come
up with dozens of theories that unify general relativity and QFT. Remarkably, the situation is the opposite, we do not have a single theory
that unifies them and is entirely satisfactory. There are several proposals, but we will concentrate on only two of them, which are pursued by
the majority of physicists working on the problem: loop quantum gravity and string theory. A good fraction of the advances in these theories
have happened in the last two decades. Loop quantum gravity stems
from a community of physicists who work or used to work in general
relativity. The main objective of this group is to unify the conceptual
frameworks of general relativity, with its dynamical space-time that is
independent of any background space-time structure, and quantum mechanics. String theory stems from a community of physicists that used
to work in particle physics. Their goal is to find a theory that includes
the Standard Model and gravity treated as another field theory. To put
it in a different way, a unification of all interactions with the framework
of QFT.

8.3.2   Loop quantum gravity

The object of study of QFT are the elementary excitations of the various forms of matter. Electrons, photons, quarks and gluons have sometimes wave and sometimes particle behavior. They are described by fields an particles. General relativity tells us that the gravitational field is embedded in the behavior of space-time. It is expected that the particle behavior in the gravitational field will manifest itself in two different ways. On the one hand there will particles analogous to photons associated with the gravitational field, they are called gravitons. On the other hand, we do not expect that the continuous behavior of space-time will hold at arbitrarily small distances. Is it therefore reasonable to think that there exist "atoms of space-time", chunks of space of irreducible volume that cannot be further divided and that in large aggregations form the space and space-time we experience, just like aggregates of ordinary atoms form ordinary macroscopic matter? Loop quantum gravity predicts a behavior of this kind. There exist ultramicroscopic building blocks of space-time.

Loop quantum gravity originated on work of Abhay Ashtekar that in 1986 proposed a description of general relativity in terms of mathematical objects similar to the ones used to treat the electromagnetic, strong and weak interactions. This allowed to study gravity with techniques that are similar to the ones used in particle physics. In particular, the use of a quantity called Wilson loop, which can be used to code all information about a space time in terms of functions of closed loops. For each loop in space the function takes a value, and if one knows its value for all loops one can reconstruct all information about the space. In 1988 Carlo Rovelli and Lee Smolin noted that loops allowed a very natural description of the quantum states of gravity. Writing the states in terms of loops quickly led to discover that areas and volumes can only take a discrete set of values, showing that at very small scales space is discontinuous, and the discontinuity arises at the Planck scale. If one considers a state of the quantum theory and computes, for instance, the volume of a region of space, one obtains a value within the allowed ones. Just like in atomic orbits, where only some energies are permitted, here only some values of the volume, area and length are. The discrete structure has an extraordinarily small scale, so these properties of space do not manifest themselves at scales we can experimentally probe.

One of the most important results of loop quantum gravity is that

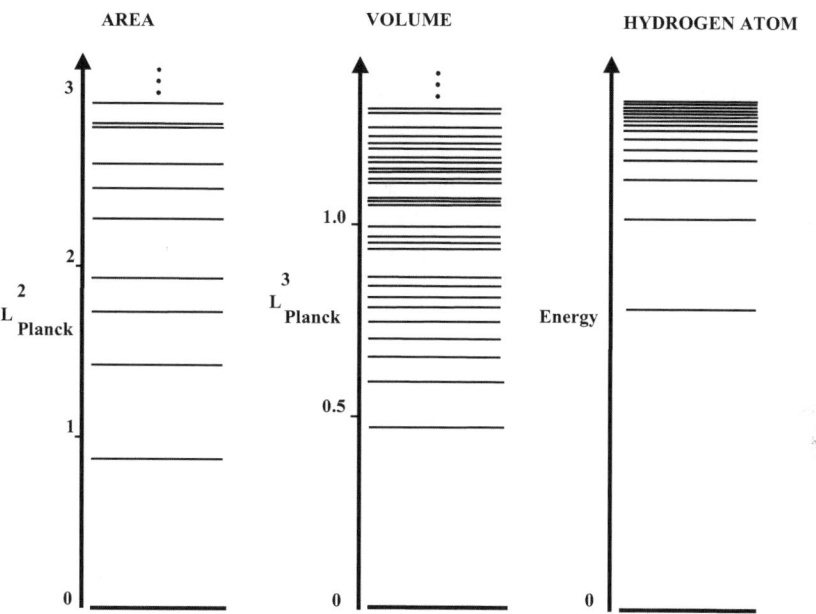

Figure 8.3: Just like the energy of an electron in a hydrogen atom only has certain permissible values, so do the values of the quantum of area and volume, reinforcing the idea of "atoms of space-time".

Thomas Thiemann showed in 1996 how to write the equivalent of the Schrödinger equation for the quantum states of gravity interacting with all the fields of the Standard Model. The remarkable result is that no infinities appear. The theory of gravity coupled to matter appears to be finite, apparently confirming the hypothesis that the divergences of ordinary QFT are due to extrapolating the idea that space is continuous to ultramicroscopic scales. The atoms of space of loop quantum gravity appear to be the cure to the divergences of QFT.

A second remarkable result of loop quantum gravity is related to the singularities that arise inside black holes and at the Big Bang in general relativity. In both situations near the singularities the curvature becomes very large. According to general relativity at the Big Bang all the universe was concentrated in a single point with infinite curvature and density. The changes to the microstructure of space-time that loop quantum gravity induces should significantly change such a picture. The state of the art calculations in loop quantum gravity concerning the Big Bang use some strong simplifying approximations. Namely it is assumed that the universe is homogeneous and isotropic and the only degree of freedom is its size. If one works out the analogous of the Schrödinger equation for the gravitational field that loop quantum gravity provides, under the assumption of homogeneity and isotropy, one finds that if one runs the universe backwards in time the densities and curvatures increase until a maximum is reached where the volume of the universe is minimum, and from there on the universe re-expands again and the densities and curvatures decrease. The Big Bang is therefore replaced by a "bounce" where the universe achieves a minimum size and then tunnels into a re-expanding universe into the past with characteristics similar to ours. If one runs things forward in time, one has a previous universe that collapses to a minimum size and then re-expands into our current universe.

In spite of these attractive predictions, loop quantum gravity is far from a complete theory. In particular it has proved difficult to show that at weak gravitational fields and low energies it reproduces the results of QFT, or for large scales it reproduces the results of general relativity. None of these limits has been achieved in satisfactory detail yet.

8.3.3   Superstrings

String theory was conceived in the 1960's as a description of the strong interactions. That initial goal was abandoned when it was realized that the theory was only consistent in either 26 space-time dimensions or 10 if one added a new symmetry, which can be viewed as an extension of the symmetries of special relativity, called supersymmetry (the resulting theory is called superstring theory). The fundamental objects of string theory are not point particles like one has in QFT but become one dimensional objects called strings. The elementary particles are excitations of the string just like the sounds of the violin are the excitations of the strings of the instrument. The natural scale of such strings is the Planck scale. The elementary excitations of the string include many particles, but they manage to include the photons, gluons, quarks and leptons of the Standard Model and also the excitations of the gravitational field, the gravitons. In 1984 Michael Green and John

Schwarz showed that there exist superstring theories that are consistent and that can incorporate a quantum version of gravity. This convinced many physicists that this was a road to unifying all interactions. It became a very active area of theoretical research. Towards 1985 there existed five different consistent superstring theories, all living in ten space-time dimensions. The main developments of this initial phase of research concentrated in perturbative calculations similar to the ones we described for QFTs. In the mid 1990's a series of exact results were obtained that showed that what people thought were five different string theories were just approximate descriptions of a single exact theory that lives in eleven space-time dimensions, called M-theory.

String theory has yielded very interesting results. The most difficult part is contact with experiment and therefore to test the theory. We live in a world with four space-time dimensions. In order to yield predictions for it, one needs to explain how an eleven dimensional theory appears to us as a four dimensional universe. The typical explanation is that the extra dimensions are very small in scale and we cannot see them, sort of like we pretend a garden hose is a one dimensional object if we look at it from far away, whereas in reality is a two dimensional cylinder. It is just that when extra dimensions "curl up" with radii that are very small one can ignore them. However, the properties of the resulting four dimensional theory depend crucially on how those extra dimensions "curl up", not just on the M-theory one started with. Unfortunately this introduces a great deal of ambiguity. The number of possible configurations that yield acceptable universes is huge, some estimates put it at the incredible number of 1 followed by 500 zeros. We could be in the unusual situation that we have a correct theory but that there is so much ambiguity in how to derive results form it that the theory loses predictive power.

8.3.4 Conclusions

Summarizing, we do not have a completely satisfactory theory of quantum gravity. We have at least two theories that in certain aspects can be considered as complementary and with certain common conceptual aspects that are worth emphasizing. Loop quantum gravity is a nonperturbative description of quantum gravity whose starting point is that space-time is background independent. Although there exist some nonperturbative results in string theory, most of its results are perturbative.

The latter require introducing a background in space-time, although it is claimed that the final results are independent of the chosen background.

In other aspects the theories coincide. To begin with, both are quantum theories. Their goal is to determine states of the system and from them to establish probabilistic predictions about various behaviors, in the end what values will be taken by various magnitudes one can measure. In that sense the quantum measurement problem is present in both theories. Additionally, both theories end up with a unification of all forms of matter: particles, fields and space-time are different manifestations of a fundamental objects, be them loops or strings. Finally, in both cases, by including space-time in the description, one ends with unexpected elements that confirm that the space-time is not just one more field. In QFT states could have complementary behaviors: they could behave like fields or like particles. Both these behaviors are present in loops and strings. The wave behavior is most manifest in the classical limit of both theories, the particle behavior arises in the quantum case with weak fields, when one considers gravitons. However, in addition to those behaviors, at very high energies one has new behaviors. In one case the discrete structure of space associated to the fundamental building blocks predicted by loop quantum gravity. In the other case the one dimensional objects of Planck scale: the strings. In both cases ordinary space-time is an emergent structure at large scales compared to the Planck scale.

Part II

Quantum Physics and emergence

Chapter 9

The problem of the interpretation of quantum mechanics and scientific realism

This chapter reproduces several paragraphs of Gambini and Pullin (2016).

9.1 Introduction

We noted in chapter 7 that the predictive successes of quantum mechanics led many physicists to not dwell on conceptual or philosophical aspects of the theory. They demanded a quick exit to problems like that of quantum measurement or the emergence of a classical world. Nevertheless, some form of interpretation was needed in order to advance in applications that appeared both in physics and in chemistry and that could not be delayed due to philosophical scruples. If the price was to forgo the construction of a coherent theoretical framework embedding a satisfactory description of reality, as we had in classical physics, it had to be done. If needed, one had to deny that physics has any other goal than to provide useful instruments to predict future observations. The very development of the Copenhagen interpretation was accompanied with operationalist points of view that showed contempt for those seeking such a theoretical framework. David Mermin summarizes this position as: "If I were forced to sum up in one sentence what the Copenhagen interpretation says to me, it would be 'Shut up and calculate!'." The point of view even reached the level of considering microscopic systems like electrons or atoms as mere formal entities destined to connect procedures to prepare experiments with the records generated at the end of them. It is worthwhile recalling the anecdote about Oppenheimer's reaction in the mid 20th century to attempts to discuss the measurement problem deviating from the Copenhagen orthodoxy mentioned in chapter 6. Or as Ghirardi mentions in his book even to take such a project seriously was taken "if not as blasphemous, at least 'unprofessional'." This position could not be held indefinitely. Due to the efforts of Einstein and Schrödinger initially, and ultimately of John Stewart Bell in the 1960's, a quest for an interpretation of the theory continued. However, up to present there is not a universally accepted interpretation. In this chapter we will highlight the need for an interpretation of quantum mechanics. We will also address up to what extent the theoretical entities can be considered as a basis to identify a realistic ontology suitable for our physical universe.

9.2 Quantum mechanics and its interpretations

Every physical theory has three elements (Jammer 1974). The first one is its mathematical framework. In quantum mechanics this is given by the space of state vectors that represent the system and Schrödinger's equation that describes the evolution of the states. We denote the mathematical framework as F. The mathematical objects must be associated to statements with empirical content. For instance, the hydrogen atom corresponds to a certain space of states. How a system is prepared, so that it is in a given state in a given instant, or how a certain device is associated with a measurement, or how to determine a probability from the wavefunction is the job of the correspondence rules, which we denote by R. The mathematical framework and the correspondence rules are the minimal content of a usable physical theory. Some physicists and positivist philosophers claim this is all one needs. Note that the correspondence rules only refer to operations and observations that will be carried out by someone studying a physical system. If we only consider F and R, physics makes no statement about the world beyond what happens in our labs. The interpretation, which we denote by I, is therefore the first conceptual model, sometimes called picture, that turns the formalism into a theory that provides a coherent viewpoint of the world. In that sense the interpretation is an integral part of the content of a theory. If one limits oneself to the mathematical framework and the correspondence rules, which only refer to phenomena in the lab, the vast majority of quantum phenomena in which the theory is relevant would be excluded of the theory. A satisfactory theory must be complete since it should make explicit predictions about events resulting from any quantum process. This is irrespective of the quantum process considered, be it a mutation that alters a DNA molecule of a living organism or a nuclear fusion process taking place in the interior of the Sun.

In this context it is worthwhile remembering the interpretational work Galileo did in order to show that the principles of the new mechanics were consistent with the heliocentric vision of Copernicus. The lesson we learn from this is that the quest for interpretation must orient itself primarily to the internal consistency of the postulates of the theory and this is precisely what is questioned by the measurement problem. What is special about the measurement processes that allows to justify going from a world of potentialities to one where there exist well defined events (the results of the measurements)? How does one explain that the evolution of the states, generically described by Schrödinger's equation, stops being applicable in the measurement processes in which states evolve abruptly? To quote Giancarlo Ghirardi, the central problem of measurement is "how, when, and under what conditions do definite macroscopic properties emerge (in accordance with our daily experience) for systems that, when all is said and done, we have no good reasons for thinking are fundamentally different from the micro-systems of which they are composed?" It is when facing the need to explain these types of issues that Dirac (1982) states that the theory must include an interpretation that allows to "extend the meaning of the word 'picture' to include any way of looking at the fundamental laws which make their self consistency obvious."

We live in an everyday world with well defined phenomena and events, and we have adapted our ordinary language to it. What are the concepts we need to use to refer to the microscopic quantum world where systems are usually in states whose properties do not manifest themselves until they are measured? Classical physics refers to phenomena to which we can always assign well defined properties and the physicist just collects data about their measurement for later analysis. Without a classical deterministic world, where the collected data can be analyzed and lead to unambiguous conclusions, our physical theories would not be viable. In the case of the quantum theory one of the central objectives must therefore be to reconcile the probabilistic nature of the theory with the uniquely defined observations that are accessible to our senses. The solution to the measurement problem should also offer an explanation of the world of events that surround us, including an explanation of how a deterministic world emerges from a probabilistic one. A good interpretation must respect the mathematical framework and the correspondence rules of the quantum theory. That may mean that F and R will perhaps suffer some changes and be substituted by another framework that opens a different world view. The potential modifications introduced must necessarily be minor since the new theory, emergent from the new interpretation, must explain the same experimental evidence explained by ordinary quantum mechanics. If this is the case one still speaks of a new interpretation of quantum mechanics, even though the latter itself may have suffered some modifications. It should be pointed out that the experimental constraints on the new theory are extraordinarily stringent. Although several alternative interpretations exist, up to present none can be considered without objections.

9.3   The derivation of an ontology from physics

Having an interpreted theory allows us to derive an ontology, that is, to recognize which are the fundamental entities that exist or may be said to exist. As Bricker (2014) puts it "typically, we accept entities into our ontology via accepting theories that are ontologically committed to those entities."

In the ontology of classical physics one assumes certain robust associations of properties and objects. These associations have certain stability with respect to time and are independent of the specific sequence of observations. For instance sufficiently frequent measurements of the position of a particle give similar values and if one decides to measure other attributes between two measurements of positions the result remains unchanged. These hypotheses, valid in any classical ontology, led to Hume's conception of the world, so ingrained in contemporary philosophy. We will return to them later on. For centuries the principles of classical mechanics as formulated by Newton and developed by Lagrange and Laplace were implicitly considered as the basis and foundation of the scientific conception of the Universe. The expectation was that the other sciences would eventually be reduced and explained in mechanical terms. Even though this goal was never achieved many areas of knowledge adopted a general mechanistic worldview. The mechanistic paradigm was superseded by relativity and quantum mechanics but it has been extremely prevalent until our days because of its simplicity and apparent consistency. The mechanistic paradigm is simple: matter is composed by elementary components (particles) which are not altered when they combine to give rise to complex structures. Classical mechanics identifies the world as a succession of instantaneous configurations of systems of material points that occupy successive positions in the mathematical space of Euclidean geometry. In this context, different phenomena produced by a system result from the different configurations that its component particles take. Given the laws of force, the motion obeys a deterministic evolution. Nothing new may occur in a classical system that is not determined by its initial configuration. *In the classical world there is causal closure, every event is the consequence of preceding events without any freedom for novelty.* The elements of the classical world are matter, the absolute space and time in which that matter moves, and the laws of force that govern movement. No other independent categories of being, such as mind, feelings or purpose are acknowledged. Cartesian dualism includes the mental aspects in terms of a new substance, being any possibility of interaction between the mental and the physical basically impossible to explain without ad hoc assumptions in this context.

The discovery of an independent form of matter as the classical fields did not change the basic foundations of the classical ontology. In particular classical fields have well defined attributes that can be measured at any time and the theory still is determinist. The notion of separability that is at the basis of Hume's doctrine of supervenience (Lewis 1986) is still perfectly justified within the context of classical physics including fields. It establishes that "The complete physical state is determined by (supervenes on) the intrinsic physical state of each space-time point (or each point like object) and the spatio-temporal relations between these points." In other words separability establishes that the total state of the Universe is determined by the states of its localized parts. As we shall see, this notion of separability no longer applies at the quantum mechanical level.

9.3.1   Quantum mechanics and the crisis in the ontology of classical physics

Quantum mechanics is usually introduced in the way we did in the first part of the book: measurements play a crucial role. Quantum systems do not have definite properties till they are measured. Between measurements their evolution is deterministic, as described by Schrödinger's equation, but in the measurements one has probabilistic behavior. One also has to assume that in the measurement processes the states change in a way not described by the Schrödinger equation. Recall the observation made by Schrödinger: if the evolution in measurement processes were given by his equation, at the end of the measurement the cat would not be dead or alive but in a superposition of both possibilities. Such presentation of quantum mechanics is known as the textbook interpretation or the Copenhagen interpretation. In fact, it is a simplification of the studies that Bohr and Heisenberg carried out to understand the microscopic world of atomic or sub-atomic scale. To be precise, Bohr and Heisenberg never totally agreed on how to interpret the mathematical framework of quantum mechanics. None of them used the term "Copenhagen interpretation" to describe their vision of the quantum theory. The term was coined later as a way to identify the points of view developed by the founding fathers of the theory and others who held the same views in the initial stages of the theory, like Max Born or Wolfgang Pauli. In spite of their differences, the ideas of Bohr and Heisenberg philosophically tilt towards a Kantian view of the world. They both certainly are physicists, so their closeness to that philosophical position is never too precise or even intentional.

Simon Saunders (2010) summarizes this situation as follows: "For despite all its obvious empirical success and fecundity, the theory was based on rules or prescriptions that seemed inherently contradictory. There never was any real agreement on these matters among the founding fathers of the theory. Bohr and latter Heisenberg in their more philosophical writings provided little more than a fig-leaf; the emperor to the eyes of realist, wore no clothes. Textbook accounts of quantum mechanics in the past half-century have by and large been operationalist. They say as little as possible about Bohr and Heisenberg philosophy or about realism"

9.3.2 Bohr interpretation

We should start by saying that the basic concepts that Bohr introduces have evolved with time and with the line of argument. His analysis of quantum mechanics is based on classical mechanics. He postulates (Bohr 1939) the correspondence argument: "... which gives expression for the exigency of upholding the use of classical concepts to the largest possible extent compatible with the quantum postulate." His reliance on classical physics seems to result from his conviction that the former conforms to our ability to gain knowledge. Resonating with Kant he seems to assume that our description of the world of phenomena happens with the aid of concepts like determinism, causality, space and time. Those concepts can only be used unambiguously in the classical world. He says (Bohr 1938): "...all description of experiences has so far been based upon the assumption already inherent in ordinary conventions of language, that it is possible to distinguish sharply between the behavior of objects and the means of observation. This assumption is not only justified by all everyday experience but even constitutes the whole basis of classical physics..." By using classical physics as a mediator between the quantum entities and the cognizant subjects, Bohr attempts to eliminate all subjective elements from the analysis. He establishes that the measurement happens in the boundary between the classical and the quantum.

Summarizing this first point, Bohr holds that the interpretation of a physical theory must be made in terms of our ordinary language that originates in our pre-scientific experience. It is framed in terms of ideas like position, duration, cause and effect. Without these notions there is no objective knowledge. Only classical physics satisfies precisely these prerequisites. That is the reason the description of measuring devices and the results of measurement must be presented in terms of ideas of classical physics: "...however far the phenomena transcend the scope of classical physical explanation, the account of all evidence must be expressed in classical terms." (Bohr 1949). Classical entities are not

altered by observation. Therefore, if one explains the probabilistic predictions of quantum mechanics through their effects on classical systems,
the observers play no role.

Bohr goes on to point out that the quantum theory appears to force
us to use incompatible classical notions to describe phenomena. For
example, the classical experiences with light suggest that we need to
treat it as a wave, but the photoelectric effect appears only compatible
with a corpuscular treatment of light. The description in classical terms
is therefore ambiguous unless one specifies completely the experiment
being carried out. Bohr (1956) says about this "Here we are clearly
in a situation where it is not longer possible to define unambiguously
attributes of physical objects independently of the way in which the
phenomena are observed." Recall for instance that the devices used
to determine the position of a particle alter its velocity uncontrollably.
Therefore the classical concepts used to describe the quantum experiences cannot all be used simultaneously. Moreover, the use of classical
concepts is really only applicable to macroscopic phenomena resulting
from the interaction of micro-systems with macroscopic measuring devices. Indeed, in order not to have ambiguities and be able to make
the probabilistic predictions that quantum mechanics allows, one needs
to know the state of the system before the measurement. That requires knowing the operations carried out to prepare the system. As a
consequence: "The essential lesson of the analysis of measurements in
quantum theory is thus the emphasis on the necessity... of taking the
whole experimental arrangement into consideration." (Bohr 1995).

Bohr admits that although macroscopic objects, when used as measuring devices, seem to behave classically, speaking precisely they are
quantum objects: "The construction and the functioning of all apparatus like diaphragms and shutters, serving to define geometry and tim-
ing of the experimental arrangements, or photographic plates used for
recording the localization of atomic objects, will depend on properties of
materials which are themselves determined by the quantum of action"
(Bohr 1948). That is, they behave quantum mechanically. In fact, Bohr
had to use the quantum properties of the measuring devices to refute
Einstein. In 1930, during the Sixth Solvay Conference, Einstein proposed a thought experiment to show that it was possible to measure
energies and times with arbitrary precision, contradicting Heisenberg's
uncertainty principle. After a few hours of reflection, Bohr found an
argument to show that Einstein was wrong. To construct it he had to

use the fact that the needle of the device that measured the energy was part of a quantum system subject to uncertainty relations in position and momenta. But then one faces a difficult problem: if the measuring devices are quantum mechanical one falls into the difficulties like the one Schrödinger confronted with his cat. How do we avoid having measuring devices that end up in a state given by a superposition of the needle pointing towards 0 and towards 1? Of course, Bohr is aware of this problem. He proposed to solve it by denying any reality to the wavefunction associated with the state of the system. His point of view is strictly instrumentalist. The wavefunction is part of an algorithm that provides "an exhaustive description of quantum phenomena in a wide area of experience." (Bohr 1956). Such algorithm can be used to predict the result of measurements in devices. The paradox in Bohr's response is that the algorithm implied by Schrödinger's equation and the wavefunction is considered useful or correct to relate classical devices that prepare the state with those that measure it, but not to describe the the devices themselves, although one admits that they are quantum in nature.

In the standard interpretation of quantum mechanics, the wave function is a representation of all our probabilistic knowledge about outcomes of possible measurements and as such is devoid of any ontological content: As Busch (2002) puts it, "In other words in the standard interpretation, the formalism of quantum mechanics or the quantum algorithm does not reflect a well defined underlying reality, but rather it constitutes only knowledge about the statistics of observed results." The classical concepts are put in doubt by this interpretation but are not substituted by better, more suitable concepts. Faye (2014) summarizes Bohr's point of view as follows: "The interpretation of a physical theory has to rely on an experimental practice. The experimental practice presupposes a certain pre-scientific practice of description, which establishes the norm for experimental measurement apparatus, and consequently what counts as scientific experience. This pre-scientific experience is grasped in terms of common categories like thing's position and change of position, duration and change of duration, and the relation of cause and effect, terms and principles that are now parts of our common language. These common categories yield the preconditions for objective knowledge, and any description of nature has to use these concepts to be objective. The concepts of classical physics are merely exact specifications of the above categories. The classical concepts... are therefore necessary in any description of the physical experience in order to understand what we are doing and to be able to communicate our results to others, in particular in the description of quantum phenomena as they present themselves in experiments; ..." But the consistency of the classical description is now in doubt because the line of separation between the quantum object and the measuring device is not the one between macroscopic instruments and microscopic objects, because even Bohr himself pointed out parts of the measuring devices need sometimes to be treated in quantum mechanical terms in order to have a consistent description of the measurement processes.

9.4　An ontology of states and events for quantum mechanics

In general, ontology is the study about the fundamental entities there are in the universe. Attempts to base the ontology on events have a long history that was reinforced by relativity and the quantum theory. In relativistic physics events are considered points in space-time. Relativity orders events: an event B is in the future of an event A if a signal may be sent from A to B. As the maximum speed of a signal is the speed of light, there are events that cannot be affected by the occurrence of A. We say that these events are causally disconnected because the occurrence of A cannot affect in any way what happens in B. In quantum mechanics, the formalism makes reference to primitive concepts like system, state, events and the properties that characterize them. The use of these concepts suggests that the theory should admit an ontology of objects and events. A quantum system is described by a Hilbert space that represents the set of its possible states and the events that may occur in the system. Based on this ontology, objects and events can be considered the building blocks of reality. Objects will be represented in the quantum formalism by systems in certain states. In an event interpretation, events are the actual entities. States describe the potentialities or dispositions of the systems for the production of certain events. The formalism of quantum mechanics associates a mathematical object called projector to each event and its properties. This is important because de fundamental elements would have a precise mathematical description. The element hydrogen is a quantum system. A particular atom of hydrogen is a system in a given state. It is an example of what we call object. It is characterized by its disposition to produce events on other systems: for instance the emission of a photon that produces a click in a photodetector.

The program of accounting for physical reality in terms of events has a long and noble tradition that goes back to Russell (1927/2007), who stated that "the enduring thing or object of common sense and the old physics must be interpreted as a world-line, a causally related sequence of events, and ... it is events and not substances that we perceive." To put it differently: for Russell an object is nothing more than a set of events that are causally connected. Although we consider this point of view a step in the right direction, we think it is incomplete for a foundation of physical reality based on events, particularly in the light of quantum mechanics. Note that for Russell an atom cannot be considered an object as long as it does not interact yielding events, whereas our definition naturally includes any concrete microscopic system as an object, given that its disposition to produce events is always defined by its state.

Concrete reality accessible to our senses is constituted by events localized in space-time. That is, by entities that occupy a small region of space-time. This was recognized by Whitehead (1925/1997) who considered that: "the event is the ultimate unit of natural occurrence." Events come with associated properties. Quantum mechanics provides probabilities for the occurrence of events and their properties. When an event happens, like in the case of the dot on a photographic plate in the double slit experiment, typically many properties are actualized. For instance, the dot may be darker on one side than the other, or may have one of many possible shapes. An association of properties with events now substitutes the postulated association between properties and objects, typical of the classical physics. Objects in a quantum universe should be understood as systems in certain dispositional state and they do not have properties until they are measured or produce events, There is only an exception to this rule: one can in principle assign some properties to pure states. These properties are the ones observed during the preparation of the state. But, contrary to what happens in classical physics, these associations are not independent of the specific sequence of observations performed on the quantum system. A measurement may destroy previous properties of the state.

The basic idea of a measurement is the occurrence of a macroscopic phenomenon, that is, of something capable of reaching perception. Thus, as noticed by Omnès, the measurement of a property of a microscopic object implies making it generate a phenomenon, in other terms, produce an event. The process of detection of photons by dissociation of silver bromide in a photographic plate leading to a cascade effect that produces the accumulation of millions of atoms of silver is an example of the production of an event: the appearance of a dot in the photographic plate. The dot and its properties have, as we have observed before, a mathematical counterpart in the formalism of quantum mechanics corresponding to projectors in the Hilbert space of the detecting plate. We are thinking in this kind of events as the building blocks of the apparent reality. Projectors characterize both the event — the appearance of a dot— and its properties. Thus quantum mechanics provides an exhaustive and mathematically precise description of events and their properties. Each primitive concept that is introduced in the axioms of quantum mechanics is associated with a mathematical concept well known in ordinary quantum mechanics, but one can only assign them a well defined philosophical meaning if one has an interpretation of the theory. For example, quantum mechanical events could not be used as the basis of a realistic ontology without a general criterion for the production of events that is independent of measurements. On the other hand, the concepts of state and system only acquire ontological value when the events also have acquired it since they are defined by their tendency or disposition to produce events.

Quantum mechanics also describes non-local systems composed by more than one particle found in an entangled state. Recall that in quantum mechanics one may have pairs or groups of particles that have interacted in the past in ways such that the quantum state of each particle cannot be described independently. A quantum state must be assigned to the system as a whole. Measurements of the physical properties of entangled particles like the spin are correlated. For instance, if one measures the spin of one of the particles oriented in one direction the other will have the opposite direction.

When one considers non-local systems like particles in entangled states, whose components occupy different positions in space-time it is not possible to speak of a state at a given time, since that is a notion that depends on the Lorentz reference frame chosen. It is not a well defined notion in a relativistic theory. However, if the state is defined by its disposition to produce events one can rigorously show (Gambini and Porto 2001) that such disposition is uniquely defined and the state in the

Heisenberg picture only changes when events take place. The disposition to produce events separated spatially in the sense of relativity, that is, not causally connected, is independent of the temporal order that one assigns to such events. In fact, the assigned order is purely conventional since it depends on the reference system used. The concept of states in quantum systems is necessarily holistic in space-time (Maudlin 2007). Very far removed from the classical notions that Einstein considered mandatory in order to do science. In one of the letters to Max Born Einstein says: "the concepts of physics relate to a real outside world, that is, ideas are established relating to things such as bodies, fields, etc., which claim a 'real existence' that is independent of the perceiving subject. ... It is further characteristic of these physical objects that they are thought of as arranged in a space-time continuum. An essential aspect of this arrangement of things in physics is that they lay claim, at a certain time, to an existence independent of one another, provided these objects 'are situated in different parts of space'. Unless one makes this kind of assumption about the independence of the existence (the 'being-thus') of objects which are far apart from one another in space which stems in the first place from everyday thinking physical thinking in the familiar sense would not be possible." Quantum mechanics has proved that this perception about reality based on classical physics that has inspired most philosophical considerations is false and that it is not required for physical thinking.

It is important to remark that having a realist interpretation of quantum mechanics not only will allows us to understand the measurement process; it also allows understanding how a world with uniquely defined properties arises from a quantum world of potentialities. Based on this ontology, objects —understood as quantum systems in given states— and events can be considered the building blocks of reality.

The event ontology we have presented has the attractive feature of eliminating the divide between the mental and the material world. As Russell (1921/2011) pointed out "if we can construct a theory for the physical world which makes its events continuous to perception, we have improved the metaphysical status of physics" According to his view we need "an interpretation of physics which gives a due place to perceptions." *An ontology of events could provide this interpretation: events in the external world are subject to a physical description while at least some events in our brain could be directly accessible as perceptions.* Both mental events and physical events would admit the same mathematical

description in terms of projectors in a Hilbert space. The main difference between both forms of events is the way we access to them: a first
person access for the mental and third person access for the physical.
As noted by David Chalmers (1989): "The distinguishing mark of the
first-person view is the air of mystery which surrounds it. This feeling
of mysteriousness has led many people to dismiss the first-person out
of hand. It perhaps has 'spiritual' connotations not unlike those of the
occult or religion. But the first-person is not to be dismissed so easily. It is indeed a glaring anomaly today, in the heyday of the scientific
world-view. If it was not for the direct experience which all of us have
of the first-person, it would seem a ridiculous concept. But it throws
up too many problems to be neatly packaged away in the kind of third-
person explanation which suffices for everything else in the scientific
world. Pity."

9.5 Interpretations that admit an event ontology

In what follows we will discuss three interpretations that admit event
ontologies. They are the Many Worlds Interpretation, the Modal Interpretation and Montevideo Interpretation. As we will see not all of them
lead to the same notion of event. As a consequence the corresponding
ontologies will exhibit minor variations depending on what interpretation is considered.

9.5.1 Events in the Many Worlds Interpretation

Everett addressed the measurement problem assuming that the wavefunction describes the Universe as a whole, including the observers, and
that it evolves continuously obeying the Schrödinger equation without
any discontinuity or collapse. In order to explain our observations he
assumed that the wave function of an observer would, in effect, bifur-
cate at each interaction of the observer with a superposed object. The
universal wave function would contain branches for every alternative
result of the measuring object. Each branch has its own copy of the observer, a copy that perceived one of those alternatives as the outcome.
Schrödinger evolution ensures that once formed, the branches do not
influence one another. Thus, each branch embarks on a different future,

independently of the others. In order to understand how the branches corresponding to different measurement outcomes become independent and each turn out looking classical one uses today the decoherence theory. For Everett, all elements of a superposition (all "branches") are "actual", "none any more real than the rest." Summarizing: Worlds "are mutually dynamically isolated structures instantiated within the quantum state, which are structurally and dynamically quasi-classical" (Saunders 2010). Let us describe the Many Worlds Interpretation thinking in a specific example, the Schrödinger cat. Recall that he considered a cat inside a box where a quantum system evolves, for instance a radioactive substance decaying. If the substance decays then it triggers the release of poison gas and kills the cat. According with the Schrödinger equation the state of the cat is in a superposition of being alive or dead. According with the Copenhagen interpretation if we measure by any means, for instance by looking inside de box what happened with the cat, then the state would suffer a sudden change. It would collapse, one either will see that the cat is dead or alive. But, again, that involves an external measurement, by an external observer, not described by quantum mechanics. Everett's Many Worlds Interpretation assumes that the wave function never collapses. In this picture both possibilities are realized. In one universe the observer sees a dead cat in the other a cat that remained alive.

The ontology of the Many Worlds Interpretation may be considered as a monist perspective where states and events have the same nature. One may consider that in this approach what we call states and events are aspects of a fundamental entity, which is the state of the complete multiverse that include al the "worlds". Notice the relevant kind of event in the Many Worlds Interpretation corresponds to events associated with observable phenomena that give rise, according to Saunders, to different "mutually dynamically isolated structures instantiated within the quantum state" . What it is assumed here is that each time an observation takes place all the possible outcomes take place, each of them in a different universe.

However, the many worlds interpretation is not free of difficulties: i) the assignment of probabilities to the different branches has found important difficulties (Saunders 2010). ii) given the fact that the state vector may be written in terms of superpositions in many different ways, in the same sense that the components of a vector depend on the system of reference used, branching is not uniquely defined. It has been argued that a mechanism called decoherence solves the problem, but decoherence only singles out a reference frame up to small deviations and the problem persists. One needs to assign reality not only to each vector in a given decomposition but to an infinite set of small variations of these vectors.

9.5.2   Events in Modal Interpretations

Modal interpretations are another approach to the measurement problem that attempts to avoid assigning a special role to the measurement process. As in the case of the Many Worlds Interpretation it is assumed that states always evolve according to the Schrodinger equation. The modal interpretation (Van Fraassen 1980) relies on the distinction between "dynamical states" and "value states" and "actual-valued" observables. The dynamical state is the usual state of quantum mechanics: it determines which properties the system may have in its future. The value state represents the physical properties that the system actually has at a given instant. One usually calls these properties instantiated properties. The Modal Interpretation assumes that physical systems at all times possess a number of well defined instantiated physical properties, these properties can be represented by the system's value state. An essential feature of this approach is therefore that a system may have a sharp value of an observable even if the dynamical state is not an eigenstate of that observable. What changes in the different versions of this interpretation is how the actual valued properties are defined. There are different versions of the Modal Interpretation. They differ in the way that the value state is defined and therefore, the observables that may take actual values. Taking again the example of the Schrödinger cat, the value state needs to be the one that at each instant of time assigns a definite property to the radioactive atom. In other words at each instant of time the value state would have the property; decayed atom and therefore dead cat or not yet decayed atom and therefore a cat still alive.

The Modal Interpretation does not assume that the state changes after a property instantiates and it assumes that the evolution obeys always the Schrödinger equation. Nevertheless it is assumed implicitly that after the observation of a property the disposition of the system in its subsequent evolution is the same as the one the system would have had if the state had collapsed. Therefore, the ontology of states and events appears to be well suited for this interpretation. However, in it the relevant kind of events corresponds to instantiations of properties both of macroscopic and microscopic systems and therefore they do not necessarily correspond to phenomena which are macroscopic by definition. We have already expressed some reservations about the Many Worlds ontology. In the case of Modal Interpretations, we also have reservations. In the first place, each of the different versions of the Modal Interpretations and the corresponding criteria for the assignment of value states to the ordinary dynamical state of the system are problematic and based on some implicit approximations which are not always satisfied. For instance in the modal-Hamiltonian interpretation the existence of isolated physical systems is assumed. A condition that in quantum mechanics is only approximately satisfied as we will discuss in detail in future chapters. Furthermore, it is assumed without a clear justification that once a value state property is observed future observations occur as if the dynamical state has collapsed even though the interpretation assumes that dynamical states never collapse. This assumption is not included in the modal interpretation postulates.

9.5.3 The (real time) Montevideo Interpretation of quantum mechanics

The closest we have to an explanation for the measurement process within the quantum theory is what is called environmental decoherence. It is based on the fact that when a quantum system interacts with an environment with an enormous number of (microscopic) degrees of freedom, the state of the quantum system suffers transitions that "almost" look like the abrupt evolutions one needs to postulate in measurements that we have discussed in some detail at the end of Chapter 6. However, even if they are difficult to detect, quantum superpositions are still present hidden in the environment and may be in principle observed. Environmental decoherence is important because the measurement processes involve interactions with macroscopic systems with many degrees of freedom that always interact with their environment. Interactions with the environment were neglected for decades and the relevance of this effect was only recognized in the 1980s. It is problematic to consider that environmental decoherence explains the measurement process because the information about the superposition of the quantum system remains hidden in the environment and may in principle be recovered.

This is contradictory with the assumption that the measuring device takes a definite value, that is has the needle pointing in a definite direction.

A new interpretation was recently proposed that explains why these hidden superpositions are destroyed. The novelty of this interpretation of quantum mechanics is the inclusion in the quantum description of another factor up to now neglected. In the standard Schrödinger description of the evolution, time is treated as a classical external parameter, but time is actually measured by physical clocks that obey quantum mechanical laws. Quantum measurements of time have a limited precision. This limitation arises from quantum fluctuations and gravitational effects like the formation of black holes and time delays and has a fundamental nature (Salecker and Wigner 1958, Ng and van Dam 1995, Frenkel 2010). This effect, first noticed more than 50 years ago has been recently confirmed by many authors. We have shown that a quantum mechanical treatment of time (Gambini *et al.* 2009) combined with fundamental limitations of measurements stemming from general relativity, lead to a modified Schrödinger evolution that allows transitions between quantum superpositions and statistical mixtures.

When one takes into account the limitations in measurement imposed by quantum mechanics and gravity, the states resulting from decoherence become indistinguishable from those produced by the measurement postulate. The "almost" of the standard approach to decoherence is removed by fundamental limitations predicted by the theory itself and the transitions from superpositions to statistical mixtures required to explain measurements are deduced. This in turn supplies an objective criterion that says when and what events may occur. Events occur when the state of a system resulting from a full quantum mechanical evolution becomes a statistical mixture (Gambini *et al.* 2010/2011/2015). The transition to the statistical mixture of the system plus environment state gives necessary and sufficient conditions for the occurrence of events. Events are assumed to occur as random choices of the system that simultaneously lead to the production of the event and the state reduction. In this interpretation it is not assumed that these choices are part of a process that the theory describes as it is the case of the Ghirardi–Rimini–Weber approach (Ghirardi 2007). The modified evolution induced by the use of a quantum time leads, for a quantum system coupled with its environment, to a state that is a statistical mixture. This provides an objective criterion for the occurrence of events and state reductions that establishes when events and changes in the state may occur without disrupting the prediction of the evolution equation. It is important to remark that within this interpretation events always occur in systems that include the interaction of a macroscopic system with a macroscopic environment leading to decoherence and therefore are macroscopic events, as the phenomena considered by Omnès which are events accessible to direct observation, for instance in a measurement device. In this sense this interpretation keeps more similarities with the Many Worlds Interpretation than with the Modal ones. Up to now one only has a precise analysis of the complete process leading to the statistical mixture for spinning particles. For other systems one can prove that the state of the microsystem coupled to the environment approaches exponentially the statistical mixture. We consider that this is enough to assume the indistinguishability. Given the fact that the distinction between an evolution that includes quantum time measurements or a quantum reduction would require an exponentially growing number of individual measurements in order to have the required statistics for detecting superpositions in the environment. Limitations referring to the existence of a finite number of physical resources in a finite observable Universe would be enough to ensure undecidability. (Gambini *et al.* 2010/2011/2015, Butterfield 2015). However, this is a point that needs further study to have a definite answer.

Summarizing, for all the interpretations that admit an event ontology and for quantum systems interacting with a macroscopic environment that has many degrees of freedom, events will be plentiful. They not only occur on measuring devices, they occur around us all the time. Measurements are nothing but the assignment of quantitative properties to events occurring in measuring devices. This interpretation leads to a complete quantum description of the universe that manifests itself through the appearance of definite events which define the constitutive facts accessible to our senses. In the rest of this book the ontology of events will play as we shall see a crucial role and implies a major departure from the dominant mechanistic view of matter.

Chapter 10

Emergence and non reductive physicalism

This chapter reproduces several paragraphs of Gambini and Pullin (2016).

10.1 Emergence in terms of an event ontology

Emergent phenomena are said to arise out of and be sustained by more basic phenomena, while at the same time exerting a "top-down" control, constraint, or some other sort of influence upon those very sustaining processes. We are interested in strong emergence, its defining characteristics are qualitative novelty and ontological non-reducibility. The notion of emergence we are proposing considers emergent entities to be genuinely novel features of the world. By definition, one talks here of causal powers that cannot be explained in terms of the micro causal powers but arise from the existence of certain macro level entities.

Strong emergence seems to be particularly relevant: if someone attempts to explain natural phenomena without denying the existence of mental processes in physical terms she must demonstrate the viability of emergence with downward causation. The central point which this concept refers to is that higher level mental events have the ability to influence the behavior of more basic levels. This requires philosophical compromises that are not easy to justify in terms of an ontology based on classical physics. In fact attempts of explanation within a mechanistic view are problematic. As was proposed by Sperry (1987) and observed by McLaughlin (1992), emergence in the context of Newtonian Mechanics needs to postulate different ad hoc, basic, configurational force laws exerted by complex structures of particles that dominate on the standard microscopic fundamental forces exerted by pairs of particles in order to explain emergent phenomena. In other words one would need forces that depend on the position and velocities of the complete system and appear only when certain values of this set of variables are taken. This should require a kind of "intelligent design" for any emergent system from a molecule to a living being incompatible with any scientific explanation. Attempts to understand emergence from classical physics are destined to fail and have led those that followed this path to believe, like Bedau (2002) does that "strong emergence starts when scientific explanation ends." It is certainly true that many classical complex systems present some forms of emergence. This is generally due to the evolution in time of nonlinear nonseparable systems. But the kind of observed emergence in classical systems is always weak in the sense that all the properties of the systems at time t are functions of the states of its components at that instant of time and correspond to what Bedau characterizes as "weak emergence".

We will show that certain quantum mechanical systems present precisely this kind of top-down control without configurational forces. It is well known that in quantum theory the physical state of a system of particles cannot always be reduced to the state of its component particles, "or to those [states] of its parts together with their spatiotemporal relations, even when the parts inhabit distinct regions of space." as noted by Maudlin (1998). We want to emphasize here that this quantum mechanical holism involves systems that have ontologically new properties and present downward causation where macro-systems have effects on their micro components (Gambini *et al.* 2015). That is, the basic tenet of strong emergentism is that at a certain level of physical complexity novel properties appear that do not result from the properties of the parts of the system or their relations and that contribute causally to the world. That is, emergent properties have new downward causal powers that are irreducible to the causal powers of the properties of their underlying base. Ontological emergentism is therefore typically committed to downward causation, that is, causation from macroscopic levels to microscopic levels. If we adopt Crane's (2001,2010) terminology our position should be considered as non reductive physicalism because it denies ontological reduction but admits explanatory reduction in the sense that the upper level properties and causal powers can be explained in quantum mechanical terms.

The traditional objections to emergence result from the explicit or implicit use of ontological concepts based on classical physics and are not tenable when assessed from a quantum mechanical ontology of events as we shall see.

10.1.1 Ontologically new properties

Let us start by showing that quantum systems may have ontologically new properties. *Quantum systems may have certain quantum states, called entangled, that have well defined properties that neither follow from the properties of parts, nor from relations among them.* To understand this statement better, let us review how entangled states are defined and contrast them with systems in classical states.

In classical physics the state of a system of particles

$$(\vec{r}_1, \vec{v}_1, \ldots, \vec{r}_N, \vec{v}_N)$$

is simply the union of the states of each particle. Its knowledge determine all the properties of the system. For instance the energy

$$E(\vec{r}_1, \vec{v}_1, \ldots, \vec{r}_N, \vec{v}_N).$$

All the properties of a classical system are functions of the properties of its components.

In quantum mechanics things are very different. Most of the properties of a system do not have well defined values until measured; for instance the position of an electron is not well defined until a dot is produced in the photographic plate and it is detected. However any quantum system in a pure state has some well defined properties. For instance, a spinning particle can take two possible values of its component along an axis $\hat{z}$: up or down. When one performs repeated measurements on particles using an Stern Gerlach apparatus on a given state one observes dots appearing in the upper region with certain probability and in the lower with the complementary probability. When the electron is in a state that leads with certainty to a dot in the upper region of the detector identified by $z > 0$, one may say that it is in the state $|z, \text{up}\rangle$. In this case one may assign the property " z up" to the state This is the only property that one can assign to this state. The measurement of any other component will not lead to a unique value: i.e. always up or always down. In general, the properties of a pure state $|\psi >$ is always associated with a set of projectors $P_1...P_N$ such that $P_i|\psi >= |\psi >$. It is only when one knows with certainty what will be the behavior of the system in certain state that one may assign it a property. In fact, as we have observed before, events have many well defined properties but typically states do not have properties until they produce events. One may assign properties to states only indirectly trough the properties observed in some events produced during the preparation process of the state.

Systems composed of several particles may also have states with some properties with well-defined values. However, these properties may refer to the system as a whole and, in these systems, there may not be any property for the states of individual particles with well-defined values. These composite systems are called entangled. More in general, entangled systems are those that have properties with well defined values than cannot be inferred from those of their constituent parts. As we will see in what follows, it might even be the case that the constituent parts have no well defined properties and yet the system as a whole does.

Consider two electrons with spin in the z direction in a state

$$|\psi_0\rangle = \frac{1}{\sqrt{2}}|1, z, \text{up}\rangle|2, z, \text{down}\rangle + \frac{1}{\sqrt{2}}|1, z, \text{down}\rangle|2, z, \text{up}\rangle. \quad (10.1)$$

Neither the state of particle 1 nor the one of particle 2 have well defined properties. No matter what component of the spin of one of the particles one measures, one has a probability $1/2$ of measuring up and $1/2$ of obtaining down. Even though each entangled electron do not have well defined properties for their spin components the total system does. For instance one can show that it has total spin $s = 1$ in Planck units , and z component of the total spin $s_z = 0$. It is only when the observations made on particle 1 and 2 are compared that one can discover the properties of the total system. One could also determine these properties when the complete system is measured. The constituents therefore now form an inseparable unit endowed with properties without the individual systems having any property —any spin component— with well defined value.

This holistic behavior is actually not an exception but is the generic behavior of composite quantum systems after an interaction. For instance, the precise vibrational modes of a molecule depend on the entangled system of electrons and nuclei. Underlying this feature is the exponential growth of states with the number of component particles in quantum mechanics in contraposition with the linear growth in classical physics. Most of these states and their corresponding properties —projectors— would have never occurred in systems with independent —non entangled— components.

Ontological novelty manifests itself in the emergence of new properties that do not result from properties of the parts. They arise only when the composite structure is constituted.

The emergent properties of such systems are crucial for explaining chemical or biological properties in physical terms. For instance: the magnetic counterpart of the properties of entangled spins work as a magnetic needle that is at the basis of navigational skills of the European robin, a migratory bird able to detect the direction and strength of the Earth magnetic field. A "sixth" sense known as magnetoreception (McFadden and Al-Khalili 2016).

The philosopher of science Paul Teller (1986) was the first in noticing that quantum phenomena show relations that do not stem from non-relational properties of their relata, as is characteristic of the classical description of the world. Entangled systems present what Teller calls: relational holism. The emergence of new properties of the whole in a quantum world where events and properties play a fundamental role is a crucial manifestation of ontological novelty.

Healey (1999) has introduced the notion of Physical Property Holism that assumes that there are physical objects "not all of whose qualitative intrinsic physical properties and relations supervene on qualitative intrinsic physical properties and relations in the supervenience basis of their basic physical parts." He observed that the existence of physical property holism in entangled systems depends on the interpretation of quantum mechanics that one adopts. Our discussion was restricted to the ontology of events and it is in this context that have proved that ontologically new properties arise.

10.1.2   Downward causation

A strong form of emergence also requires downward causation, namely, the emergence of novel causal powers. Here a double goal arises: to characterize such form of causality in physical terms and to show that at least certain systems, like the quantum ones, exhibit downward causation.

A notion of causality that is suitable to the ontology of states and events has been developed by Chakravartty (2007). He founds his notion on what he calls "causal properties". As we have associated the notion of property to projectors in the Hilbert space we prefer to speak about causal powers. Following Chakravartty we will recognize a causal power for its capacity to confer dispositions on the objects that have them to behave in certain ways when in the presence or absence of other objects with causal powers of their own. This dispositional idea of causality —originally due to Heisenberg— is the one we have adopted in our work for states: recall that quantum states characterizes dispositions to produce events in their interaction with other systems.

Causality is normally presented in terms of related events. In quantum mechanics this vision is incomplete if the concept of state is not included. Indeed, among the events that prepare a state and the ones observed usually there is a period of time and the disposition to produce events is not given just by the initial preparation. It is indeed given by the state. The latter is defined in terms of initially observed events and their time evolution. Changes in the state are determined by its Hamiltonian evolution. Only the state at a time t defines completely the causes that lead to the observation of events in t. Quantum mechanics introduces a probabilistic notion of causation that involves states as causes and events as effects.

An object will present downward causation if the parts have some behaviors that are dictated by the state of the whole and that cannot be predicted from the knowledge of the state of the parts. The previous example of an entangled state shows that in quantum mechanics there is state non-separability (Healey 1999). In the example of entangled spinning particles the states of the parts are represented by reduced density matrices and they are just a statistical mixture of "up" components and "down" components while the complete entangled state has more information, in particular information about correlation between the events observed in each component. For instance the spin measurements of both particles will be correlated. Whenever both observers measure the spin in the same direction their results would be opposite, —up for Alice and down for Bob or vice versa—. The observers could never have figured out these correlations by looking at the individual systems in isolation without comparing their measurements. The complete system has certain non locality such that when one electron chose to answer up, the other necessarily needs to chose down. Such correlation does not involve time. As it is well known Bell's theorem establishes that it is not possible to explain this kind of behavior assuming that each part follows a pre-established set of instructions, in other words, assuming that each part has some local hidden information telling it how to act before each measurement. In other terms the state of the entangled pair of particles given for instance by $|s = 1, s_z = 0\rangle$ is not mathematically determined by the the states of the parts. In fact there are two states $|s = 1, s_z = 0\rangle$ and $|s = 0, s_z = 0\rangle$ whose parts are in identical states.

These correlations are generic for any entangled state. Any system in an entangled state presents downward causation. The events produced by different components of the system have correlations that do not result from the states of the parts. In fact, the very characterization of the entanglement of a pure state in terms of the Von Neumann entropy or entanglement entropy that is always a positive quantity shows that the complete information required for the determination of the whole cannot be recovered from the information of the parts or that the state of the whole cannot be mathematically expressed in terms of the states of its parts. The states' roles in causation, together with their non-separability when they are entangled, imply downward causation.

Less trivial explicit examples of downward causation may be found in the molecular behavior where entanglement cannot be ignored or in quantum computers. In quantum chemistry calculations, entanglement is related with the correlation energy. This energy is neglected in Hartree–Fock calculations and the energy error of the Hartree–Fock approximation to the wavefunctions of a molecular system is a measure of the effects of downward causation in the behavior of the molecular components. For instance, it is downward causation of the whole molecular system state that determines the precise vibrational behavior of the nuclei. At a first approximation, the behavior of the nuclei may be described by its motion under and effective potential created by the average behavior of the electrons. Most calculations are based on this type of approximations, but one actually has a complete interacting system of nuclei and electrons, and the precise behavior of the system depends on electromagnetic interactions that are a function of the orbital and spinning properties of all the components.

For quantum computers, the existence of quantum correlations in the entangled states between different input and output outcomes is at the basis of the application of quantum algorithm that allow to solve certain problems like integer factorization using Shor's algorithm much more quickly than any classical computers. The disposition of the quantum computer to produce the correct correlation between the input and output results at the end of the computation is the manifestation of the quantum downward causation that is at the basis of the improvement of the computational capabilities of quantum computers.

We have characterized the downward causation of a system by its disposition to produce certain effects that is not present in the dispositions of its parts. This kind of disposition is the characteristic feature of systems in entangled states. In turn, the existence of entangled states in quantum systems result from the exponential growth of Hilbert spaces of composed systems in opposition to the linear growth of the states in classical physics. At the basis of the novelty and non-reducibility of emergent systems it is this exponential growth in the possible behaviors of the quantum systems whose philosophical implications can be recognized when the appropriate ontology is put into action.

Summarizing, emergence is not the exception but the rule in interacting quantum systems, its so natural that it usually remains unnoticed. Strong emergence manifest itself in most complex systems, it is a natural result of the interaction of quantum objects.

The recent advances in the understanding of the role of quantum mechanics in biology (Vedral 2011) and the previous analysis of strong emergence in quantum mechanics rise the expectations of understanding mental phenomena and their causal powers in physical terms. The issue of emergence, also known as non reductive physicalism in the context of studies of the mind brain problem, has been extensively analyzed mostly using notions of supervenience that assume some form of separability and are not valid in quantum mechanics as shown by Maudlin (2007). We consider that in quantum entangled systems with downward causation and dispositional states that leads to probabilistic outcomes the issue of free will may be posed in a clear explicit way and will be analyzed latter on.

10.1.3 Summary

Several interpretations of quantum mechanics admit event ontologies. These realistic interpretations lead to an important revision of the notion of matter and its potentialities. Systems of particles in entangled states have new behaviors and emergent properties.

The quantum theory implies that the lower levels are modified even up to the point where they lose part of their individuality when they integrate into an entangled system in a higher level of the hierarchy. The emergent structure has novel properties and downward causation. Interpretations of quantum mechanics that admit an event ontology solve the traditional problem of explaining emergence. We have analyzed several examples of emergent systems as molecules or the chemical bond that emerge from physics. We have mostly restricted to physical and chemical results because they admit a simple analysis but emergence due to a similar mechanism where the state of wholes and their causation are not determined by their parts are also present in biological systems as we will discuss with more detail in the following chapter. The relevant result is that there is a hierarchy of structures where the top levels influence what happens in the lower ones with top-down causal powers. As George Ellis (2016) observes: "A full scientific view of the world must recognize this, or else it will ignore important aspects of causation in the real world and so will give a causally incomplete view of things. Complex causation is based in the interaction of bottom-up and top-down effects: if we neglect either, we will be unable to understand genuinely complex systems" As we shall see, it is due of emergence that higher level mental functions are not epiphenomena, as Ellis puts it, "they are causally effective autonomous entities in their own right, even though are made possible by the underlying physical states."

Part III

The centrality of life

Chapter 11

Darwinism and the centrality of life

11.1   The issue of the origin of life

A completely satisfactory definition of life would imply an understanding of life phenomena that we currently do not have. Any description is therefore provisional and can be subject to criticism. For a biologist, life is the behavior of certain open physical systems capable of interchanging matter with the environment and that present homeostasis and metabolic activity. That is, they are capable of self-regulation in such a way as to maintain stable dynamical equilibrium conditions transforming matter and energy from their exterior through a self-sustaining internal process that involves the production of its own components. In addition to fulfilling this fundamental criterion, without which no living organism can exist, other distinctive characteristics of living organisms are: their internal organization, present at the cell or multicellular level; their ability to adapt to the variations of the medium they live in and therefore to respond to stimuli from that medium; their reproductive capacity, that is, their ability to produce new similar organisms either asexually from one mother organism or sexually.

The problem of the origin of life on Earth has proved very difficult to attack with the current scientific tools. There is evidence that suggests that life appeared on Earth some 3,700 millions of years ago, or even

before. This takes into account that the Earth is 4,500 million years old
and that it is unlikely that oceans formed long before 4,000 million years
ago. But the formation process for the oceans and atmosphere is subject
to great uncertainties. We only know that the current atmosphere with
its high content of oxygen is the result of biological action by bacteria
and its formation appears to have taken place some 2.9 and 2.4 billion
years ago.

Given the experimental evidence acquired recently about the existence of biogenic graphite in Greenland, it is assumed that some 3,700
million years ago there already existed some forms of life. Since all
known life shares a set of basic common properties like the genetic code,
it is generally assumed that life stems from a common ancestor called
the last universal common ancestor (LUCA) of all living things presently
on Earth. LUCA should not be assumed to be the first living organism,
there were probably other early microbes. However, the fact that many
more amino acids are possible than the twenty present in modern life
points to a common ancestor. Although this common ancestor should
have been very simple as compared to present living systems, in order to
be able to evolve one must assume it had complete blueprints for DNA
replication, protein synthesis and RNA transcription. A system of this
nature with genes and proteins working together does not seem to have
evolved from what could be a boiling cauldron of chemicals that formed
on Earth after the creation of the planet about 4.6 billion years ago, unless a series of yet unknown physicochemical mechanisms had somehow
controlled the process. If this chronology is correct and life originated
some 3,700 million years ago it could be said that life on Earth arose
"as soon as it could", which in turn seems to imply that the origin of
life was based on robust reactions.

Biological functions are regulated by the catalytic power of enzymes
and the ability of nucleic acids to store information. In order to work,
very large molecules with long chains of components known as residues
are needed, for instance aminoacids. Molecules like proteins with long
chains allow when diluted in water to have many active residues located
close to each other thanks to the folding of the molecules into characteristic shapes and with certain resistance to changes of architecture.
Basically, the long chains allow the organization of the space surrounding the molecule giving rise to a micro environment with well defined
physicochemical properties. It is not clear how the proteins that constituted LUCA were selected from the whole set of possible proteins that

is an enormous number. For example there are 20^{100} proteins with 100 aminoacids, whereas the number of different proteins that exist in our planet is around a billion. What is the selection mechanism? There are several possibilities. One we cannot rule out is that of all those possibilities only a small number are energetically allowed. Another is that the special conditions that exist on Earth could only lead to a restricted number of molecules and that these molecules resulting from a merely contingent process contained the adequate ones to start life. The final possibility is to resort to the *anthropic principle* and claims that since life exists, the right proteins had to be selected no matter how low the probability. We therefore see that the probability of creating a living organism by mere chance does not seem to allow, without a virtually limitless number of trials, to have an explanation. We therefore need physicochemical mechanisms that favor the emergence of life if life appeared so soon in the evolution of the Earth. An example would be the aforementioned energetic restriction in the number of proteins that are physically realizable. Some recent advances go in the same direction by showing that there exists a strong correlation between the physical properties of aminoacids, particularly concerning their behavior in aqueous media with hydrophobic properties, and the form in which proteins fold when diluted in water and the genetic material contained in the RNA. And that those behaviors persist at the high temperatures that one assumes dominated when life originated.

In recent decades important discoveries have been made concerning the enormous redundances that exist in possible metabolisms performing the same function, or different proteins able to produce similar effects (Wagner 2015). This allows evolution to explore different behaviors without loosing the viability of the organism. At the same time, totally new capabilities can be achieved with a small number of mutations. These properties of biological information storage are what endow organisms with the ability to adapt to an ever changing world. One way of achieving this goal is allowing a great number of proteins to carry out the same functions. In the case of the production of ATP, the basic energetic biomolecule of the cellular metabolism, Keefe and Szostak (2001) showed that more than 10^{93} proteins can bind ATP. This in fact poses a new, perhaps more important, problem. Natural selection in not able to innovate, it only allows innovations to spread; it is the complexity of the molecular design of proteins and genetic material which allows life to adapt to an enormous number of possible environments. Thus, most of the plasticity of life was already present somehow in the LUCA.

It appears that although life on Earth started very soon and as a consequence there must be physicochemical restrictions that favor the formation of the required structures, that does not imply the process was inevitable. Very likely in addition to physicochemical mechanisms, chance played a role at least in the selection of the proteins involved and in the form of codifying the information.

A lot of people facing the choice of contingency and determinism clearly choose contingency. Even though this would make the origin of life an absolutely exceptional process in the Universe. The explicit motivation is to avoid any suspiciousness of the existence of something resembling purpose in the behavior of physical laws. For instance, P.L. Luisi (2016) states that "Can one say a final word about this dichotomy contingency/determinism? It would be wise, of course, to avoid the extremes and look for a balance. The image that comes to mind is one used by Maturana and Varela (1998), when discussing the subject of biological evolution; consistently with Kimura's views on evolution, they use the metaphor of water falling from the top of the hill: the flow of water is determined by gravity, by the laws of nature. However the actual path is determined by the accidents on the ground —the trees, grooves, and the rocks encountered on the way, so that the actual downhill flow of water is a balance between the forces of determinism and contingency. Compromises like this are always useful and make life easier. However, often they fail in the most critical situations. For example, take one fundamental question in the origin of life: is there a transcendental power behind it, or not? It would be nice to find a balance, a hybrid between Scylla and Charybdis, but, unfortunately, this is an either/or situation." From a strictly scientific point of view we believe it is clear that without mechanisms induced by physical laws the phenomenon will be equally incomprehensible either if we attribute it to a highly unlikely accident or to a divine creation. Although with time processes may be discovered that favor the development of life, we are likely to have a definitive answer to the question of the exceptionality of life only when exobiology allows to determine with what frequency life is found in extrasolar planets that orbit in regions potentially inhabitable.

Part of the problems stemming from the origin of life are due to the dominant approach of biological evolution based on natural selection being inapplicable to prebiotic process. Therefore natural selection cannot account for the processes that made it possible. As a scientific practice,

Darwinism has transformed into a position that excludes other explanatory frameworks and that tends not to take into account other phenomena, like self-organization, that are present in evolutionary processes and
surely played a key role in the origins. As Wimsatt (1998) says "one does
not need special circumstances —or selection— to form self-organized
state or properties: one needs special circumstances to prevent them."
As in the previously mentioned examples about the way in that proteins
fold due to the properties of aminoacids, self-organization emerges from
processes in which properties of the global system (the way the protein
folds in water) emerge from multiple interactions between the lower level
components (the aminoacids) that constitute it.

11.2 Darwinism

The Darwinist explanation of the evolution of species was initially formulated in the first edition of "On the origin of species" in 1859. Charles
Darwin was born in 1809 and lived in an environment rich with political and religious illuminist ideas. His thought was strongly influenced
from the ideas that were beginning to spread about geology and natural
history. Two authors were decisively influential in the scientific development of Darwin: Sir John Herschel with his book "Introduction to
the study of natural philosophy" and Charles Lyell with his "Principles
of geology". Reading and re-reading this book helped him pass the time
during his trip around the world of almost five years aboard the Beagle.
As Darwin himself admitted, Lyell's ideas were profoundly influential
on his thinking. Lyell considered that the main purpose of geology was
to register the changes that had taken place on Earth and understand
their causes in terms of processes that are still operating in our time.
One of his main observations was the discovery of a constant process of
appearance and extinction of species that had to be explained in terms
of the geological phenomena that accompany such changes.

   With those influences and with a meticulous field observation col-
lecting specimens and careful notes about the observed species he starts
to develop his ideas about the origin of species. This problem had been
cataloged by Herschel as "mystery of mysteries" of natural history. He is
soon convinced that a natural explanation of the processes had to imply
a gradual transformation of species whose causes required identification
in a naturalist scheme. Darwin's explanation of evolution of species can

be briefly stated: 1. Plants or animals that constitute a species have a tendency to reproduce having more offspring than those that can survive since resources are limited, creating a struggle for survival among the members of the species. 2. The members of a species show small variations among themselves in their characteristic features, making some better adapted to the environment in which they live. They latter will tend to survive more or leave more offspring, gradually replacing the less adapted, since the offspring tend to inherit the favorable variations of their parents. 3. With time the small variations accumulate and new species will arise. 4. If the selection takes place in different ways in different regions with different environments, from a single species many new ones can arise. This process can repeat indefinitely.

A key element of the natural selection process, as described by Mayr (1998) is "the nonrandom and reproductive success of a small percentage of the individuals of a population due to their possession, at that moment, of characters that enhance their ability to survive and reproduce."

Darwin's conception of evolution had a strong impact on the traditional image of mankind as center and final objective of creation. The traditional image had even survived the Cartesian analysis that attempted to consider all living beings —all animals, with the exclusion of men— as pure mechanical systems. Nevertheless Descartes held the notion that mankind had a protagonic role, being the only creatures endowed with intelligence. Descartes had reserved one of his two substances to describe the mental, excluding the animals, which were conceived as mere mechanisms. The high esteem in which mankind held itself received several blows. The first one was the discoveries stemming from Newtonian physics and the verification of Copernicus' heliocentrism, which took the Earth away from the center of the Universe. It was just a stone traveling through an incommensurate, almost empty, space. The second blow was Darwin's discovery that mankind is just another species taking part in evolution. These developments are at the root of the nihilist movements that appear at the end of the 19th century. Friedrich Nietzsche (Molina Y Vedia and Grimm 1995) summarized the situation as follows: "Has the self-belittlement of man, his will to self-belittlement, not progressed irresistibly since Copernicus? Alas, the faith in the dignity and uniqueness of man, in his irreplaceability in the great chain of being, is a thing of the past —he has become an animal, literally and without reservation or qualification, he who was, according to his old faith, almost God ("child of God," "God-man"). Since Copernicus, man seems to have got himself on an inclined plane —now he is slipping faster and faster away from the center into— what? into nothingness?"

The tendency to devalue the privileged image that mankind had of itself becomes emphasized with The Origin of Species. It is a process that deepens with Freudian psychoanalysis and the attempts to explain mental phenomena in mechanist terms. In biology progress goes in the same direction with the neo-Darwinian vision of the selfish gene that attempts to eliminate the living organism from the evolutionary drama by substituting it by a mechanical process of preservation of the genetic information contained in the organisms. Far from being definitive scientific developments it appears that the theories that arose in the previous century result from an effort of embracing in a dogmatic way a mechanist vision that is a priori considered the only one compatible with natural sciences. However, with the development of quantum physics in the 1920's a reverse process develops. In it, the impoverished conception of matter of classical physics, which had already started to change with the introduction of the notion of fields, is transformed and enriched further to the point of being unrecognizable and very difficult to accept. This inverse process that as we shall see could be called humanization of matter has been consolidating till today. In particular, thanks to recent developments in biology resulting from the genome and the incipient development of quantum biology. Our main objective in what follows will be to travel this inverse road and analyze its consequences. To capture it as a slogan: the reduction of spiritual to material is followed by the transmutation of the material into spiritual.

11.3 From neo-Darwinism to evolution in the era of genome

Darwin's theory has many limitations that have to be recognized and have been pointed out throughout the years. The most obvious ones are, firstly, that it lacks predictive power since environment changes cannot in general be predicted. At best it allows to explain why some species have originated in the changes that took place. Another important limitation is that the observed changes in nature do not depend only on natural selection and many times derive from the structural or biological characteristics of the species involved. A third source of debates with exchanges of arguments and examples in favor of the various positions is the one of the so called *units of selection.* Assuming that evolution is the result of natural selection, what gets selected in the end: is it the species, the organism or the genes? Although it may seem a somewhat technical debate, behind it lurks the conflict between the points of view outlined at the end of the last section between a mechanist reductionism on one hand and an organicist holism on the other.

The development of genetics during the 20th century gave rise to a version of Darwinism that became dominant in the second half of the century. It has received different denominations, some clearly pejorative. We prefer the most neutral one: "neo-Darwinism". It most notable defender in the last decades has been Richard Dawkins, author of the book "The God delusion" to which we have referred already. According to Dawkins (1976/2006) "A gene is any portion of chromosomal material that potentially lasts for enough generations to serve as a unit of natural selection." In terms of genes, evolution is understood as consisting in the change of genetic constitution of a population. A gene is favored by the natural selection process if the frequency with which it appears in a certain population grows with time. The theory distinguishes the unit that contains the information contained in the gene, called replicator, from its vehicle, that is the organism that interacts ("interactor") with the environment and is selected for its ability to proliferate in it. The units that in the end are selected are the genes —the replicators—. According to Dawkins replicators "...are in you and me; they created us, body and mind; and their preservation is the ultimate rational for our existence. They have come a long way, those replicators. Now they go by the name of genes, and we are their survival machines." Neo-Darwinism substitutes the organism by the gene. Evolution takes place at genetic level in a process that involves random variations and gene selection. As Lenny Moss observes, neo-Darwinism "denies agency to living organisms in favor of the abstract dictates of an algorithm or the logic of 'the replicator."'

In the last decade with the advances in the understanding of the genome this position has received numerous criticisms (Conway Morris 2004, Moss 2004). According to Moss "When scientists and clinicians speak of genes for breast cancer, or genes for cystic fibrosis, or genes for blue eyes, they are referring to a sense of gene defined by its relationship to a phenotype (i.e. the characteristics of a person or whole organism) and not to a molecular sequence. The condition for having a gene for blue eyes or a gene for cystic fibrosis entails, not a specific nucleic acid (DNA) sequence, but rather the ability to predict,...., the likelihood of some phenotypic trait." DNA sequences are not directly related to characteristics of the organism but with its capacity of producing certain proteins. It is the presence or absence of these proteins, that depends on environmental factors in additions to the ones coded in DNA, which ends —among other possible effects— determining the characteristics of the organism (its phenotype). In its most recent definition, the latter means: set of visible characteristics that an individual presents as a result of the interaction of its genotype and the environment. Moss has proposed distinguishing between the P-gene, associated to the visible characteristics (the phenotype) and the D-gene associated to the molecular sequences that are present in the DNA.

Neo-Darwinisms central hypothesis is that the hereditary variations depend only on chance and are independent from the interactions of the organism with the environment. That implies assuming two hypotheses: first that the genetic material is absolutely independent of the environment and the life development of the organism and second that the changes in that material have direct effects on the organism independently of other factors like the environment.

The first hypothesis assumes that the genetic material is protected from influences stemming from the development of the organism. This hypothesis was formulated by Weismann at the end of the 19th century and today it is known that it is not applicable to a good fraction of living organisms, although it does to most vertebrates. It is doubtful that a characteristic that appears acquired during the evolutionary process would be the basis of such process. But we will provisionally admit that the changes in the genetic material are due solely to spontaneous mutations, contrary to the arguments of many authors in the last decades (Sapp 2003, Radman, Matic and Tadei 1999).

The second assumes that variations in the ADN sequence directly determine the phenotypical characteristics of the organism. This is crucial for natural selection to act favoring or disfavoring those variations. It is in this hypothesis that discoveries of the last decades are most clear in denying a direct relation between changes in DNA and the phenotypical characteristics of the organism. The most spectacular evidence of this point, as Moss points out, arises from the capability of producing transgenic animals where some molecular sequence that has been correlated with certain functions has been eliminated. For instance it is possible to eliminate completely a gene that when damaged has manifest carcinogenic effects. Eliminating this gene, far from enhancing the damages, produces organisms that look perfectly normal. Attempts to explain this phenomenon alluding to the existence of redundancies in the genes have become less and less viable since it was discovered that the number of genes is not correlated with the complexity of an organism.

If the study of genome in the last decades has shown something is that there does not exist a correlation between the number of genes and the complexity of an organism. As Lynch (2007) observes "One of the greatest surprises of the Human Genome Project was the discordance between the count of protein-coding genes ([about]) 24,000 [for humans]) and expectations based on perceived phenotypic and behavioral complexity." The estimate of the number of human genes has been progressively reduced from initial estimations of 100,000 or more as genome sequence quality and gene finding methods have improved and appears to continue to reduce. Some recent estimates put it at less than 20,000. As Cooper (2000) says, "The genomes of salamanders and lilies contain more than ten times the amount of DNA that is in the human genome."

It has been understood that the complexity of the human genome, that is the genetic information contained in chromosomes, is not rooted in the number of genes. It rather depends on how part of those genes are used to construct different products in a process that is called *alternative splicing* that allows that a single gen codify many proteins. The specific protein produced in each case depends on the needs of the organism and its environment.

The match of the response of the genome to the demands of the organism is clearly shown in the differentiation of the embryo's cells. It is apparently contradictory to attempt to explain the embryo's differentiation in terms of the behavior of genes that are identical for all the cells of the organism. To explain the development in purely genetic terms, how do identical genes give rise to such differences?

Modern biology has found an explanation for this behavior in epigenetic terms. Epigenetics deals with the regulation of the transcription. Its first modern definition was by Conrad Waddington and it implied the study of the interaction between the genes and their products that yield a phenotype. In more modern terms, if the DNA contains the necessary information for the expression of all proteins and RNAs needed for the signaling, synthesis and organization of all the substances needed for the survival of a living organism, epigenetics deals with the factors that modulate the expression of such genes. It therefore deals with a great number of modifications, with widely varied mechanisms of action. The most studied ones are the methylation of DNA, the covalent and non-covalent changes of the histones and the non-coding RNA. Summarizing, epigenetic regulation determines not only what parts of the genome are expressed but also how they are spliced.

As a summary we can quote Moss (op. cit.): "The evolution of increasingly complex organisms, it turns out, is based upon the evolution of increasingly modular architectures. The critical decisions made at the nodal points of organismic development and organismic life are not made by a prewritten script, program or master plan but rather are made on the spot by an ad hoc committee. And these committees consist of ensembles of modular parts, the composition of which are contingent upon circumstance. And the more complex the organism, the grater the number of different potentially modular constituents and the more sensitive is the outcome to the nuances of circumstance." That individual cells can be capable of coordinating such a large set of decisions that determine their development is something admirable that we are still far from understanding.

Summarizing, the organisms, far from playing a passive role in the evolutionary process are the main protagonists of it. The creative process of activation and deactivation of potential capabilities in a genome that admits a countless number of possible combinations, resides in the organism. In the end a portion of the DNA only acquires the function of a gene inasmuch as as it is activated by the upper levels —that is, by the organism and in many cases by the demands of its environment— in a process of downward causation. The genetic information cannot be identified by its effects on the phenotype but by its role in a "genetic toolkit" that is used in different ways according to the needs of the organism. Another analogy for the oversimplified vision of the neo-Darwinist standpoint that seems suitable to the situation would be one that identifies the culture of a civilization with the information preserved in its written documents, its books.

As an example of the recent discoveries concerning the genome and its indirect effect on phenotypical development it was recently pointed out (Agaba *et al.* 2016) that changes in a few dozens of genes explain the unusual development of the giraffe. Sequencing its genome allowed to understand how that animal evolved into the tallest on the planet. The genetic sequence of the giraffe was compared with that of the okapi, with whom it shares a recent common ancestor of just 12 million years. It was verified that although both share the majority of the slightly over 17,000 genes in their genome, in 70 of them the giraffe accumulates a series of changes that can explain the evolutionary success of an animal that can reach 20 feet in height. Such height requires a large heart to pump blood six feet up the neck and strengthened blood vessels in the lower extremities to handle the resulting pressure. At the same time its digestive system has learned to take full advantage of the toxic acacia trees they ingest. The aforementioned genes have information about certain proteins that control cardiac development like the skeletal muscle, which suggests that the long neck and long front legs evolved in parallel with the heart and the circulatory system. Important modifications were also found in factors affecting growth. Genetic changes appear to have accompanied environmental and nutritional changes in a process that combines the bottom-up with the top-down causation.

11.4   Can one talk about progress in evolution?

There is no doubt that the process of evolution through natural selection implies, as Darwin himself admits, a process of exploration of possibilities larger and larger. Darwin (1859/1999) comments: "It may be said that natural selection is daily and hourly scrutinizing, throughout the world, every variation, even the slightest; rejecting that which is bad, preserving and adding up all that is good; silently and insensibly working, whenever and wherever opportunity offers, at the improvement of each organic being in relation to its organic and inorganic conditions of life." It is therefore not surprising that in the process of exploration of possibilities the competition imposes a selective pressure, almost like in a gas diffusion process that leads to the occupation of niches in the biosphere that are still not occupied. This process sometimes is called horizontal (e.g. Wandschneider 2005) and is not connected with the development of new or superior aptitudes but with the exploitation of new environments free of competitors. Among the organisms that develop this type of strategy are the extremophiles, that is, organisms that thrive in extreme conditions that are detrimental to most life on Earth.

That is how one finds organisms for whom the optimal conditions of life are chemically very acidic or basic or live in extreme temperatures.

Wandschneider distinguishes a second form of evolutions that he calls vertical. He says: " Horizontal evolution consists of the occupation of available biospheres, but vertical evolution creates new biospheres. For example, the existence of plants makes the existence of herbivores possible; the existence of herbivores, in turn, makes the existence of carnivores possible. In the first case the botanical world created by evolution provides a food resource, but only for a completely different kind of living being, namely herbivores, and these in turn provide a food resource for carnivores." This process gives rise to a series of levels each of which presents a completely new set of functions. Several categories of new functions appear in herbivore animals: mastication and digestion, motility, perception and emotion. Digestion allows the animal to process the food provided by the plants without the need of chlorophyll or photosynthesis. The independence of a particular source of mineral nutrients allows motility and the ability to move around requires the ability of control of the motor system, a sensory organization, information processing and a brain. The identification of a food source that may imply directed long-range motility requires as Jonas noticed "not only developed motor and sensor faculties but also distinct powers of emotion. Greed is at the bottom of the chase, fear at the bottom of flight."

In vertical evolution what has been achieved in a certain stage is accessible in the next one, maybe indirectly through the lower level that gets incorporated into the environment of the higher level. For instance the clorophyllic function is not needed when its products are accessible with the development of new forms of nourishing. The acquisition of new biospheres does not appear to be a lucky peculiarity of terrestrial evolution but the result of the exploration of new environments intrinsic to the permanent search of alternatives that allow survival. This vertical evolution mechanism would allow to understand what naturalist and sociobiologist Edward O. Wilson (1999) observes: "During the past billion years, animals as a whole evolved upward in body size, feeding and defensive techniques, brain and behavioral complexity, social organization, and precision of environmental control." These, among others, are the changes that characterize progress in biology.

Perhaps what is most paradoxical is the permanent emergence of self-organized phenomena that are the result of physical laws and that act limiting the number of possibilities that can be generated by mere chance. In this sense evolution does not result solely from natural selection but from a process of selection among structures allowed by the laws of physics. For instance the ways in which a protein can fold in order to give rise to a specific spatial distribution with certain functional roles in the organism are determined by the physical laws and therefore pre-existed as possibilities as modular structures on which vital phenomena can be built. These forms of self-organization are, on the other hand, the only mechanism that could have guided the processes that gave rise to the first forms of cellular life in stages in which the process of natural selection could not have taken place since the genetic material, DNA and RNA —crucial for its development—, had not been constituted yet.

11.5   Mechanism, Darwinism and 21st century science

With the birth of modern physics with Galileo and Newton a concept of matter deprived of all the characteristics of life takes center stage. Let us recall that Galileo was who first suggested a distinction between what would be later called the primary intrinsic properties of an object, among which are the concepts of space and time we just mentioned, and the secondary ones associated to the sensations of color, smell, taste or sound that designate changing states of our mind. He says "let us mentally suppress the living beings and their organs and these qualities disappear from the world."

This is the essence of the mechanist position that strips reality from any attribute that is not mathematical or at least mathematizable by the science of motion. Alfred North Whitehead (1925/1997) describes this position eloquently when he states that "Thus nature gets credit which should in truth be reserved for ourselves: the rose for its scent, the nightingale for his song, and the sun for his radiance. The poets are entirely mistaken. They should address their lyrics to themselves, and should turn them into odes of self-congratulation on the excellency of human mind. Nature is a dull affair, soundless, scentless, colorless, only the hurrying of material, endlessly, meaninglessly."

The Aristotelian universe, limited by the celestial sphere with Earth and mankind in its center is transformed into an abyss of limitless space and almost empty in which inanimate forces act on material bodies that move obeying the laws of inertia. The reduction to the exact knowledge was achieved reducing the material to a mere extension and elevating the mechanism devoid of life to everything that was knowable.

That was the Cartesian program, started by applying the principles of the new physics to the most general possible setting. Not only the planets and the inanimate objects will obey the strict deterministic laws of mechanics, also the plants and animals are treated in mechanist terms. Human physiology receives a similar treatment and in that sense Descartes sets out to propose different mechanical models for bodily functions like eyesight. Only the realm of the mind, reserved to man, the author of that feat of reason that is the new physics, is excluded. Thus he constructs the world starting with two substances with distinct and incompatible attributes: the bodily substance and the mental substance. Indeed if one describes the operation of the senses and the brain in purely mechanistic terms, as if they were clockworks, the connection between a state of the body and mind turns into a mystery.

The very existence of life in a mechanical universe gets transformed progressively into a problem that must be explained and that in fact does not admit any possible explanation without in turn transforming the way of conceiving the physical world. Today we know that even to explain the chemical processes and the existence of atoms and molecules it is crucial to use quantum mechanics that gives rise to an enormously richer vision of matter. But even today the essential problem of Cartesian mechanism persists: the existence of sentient life in an inanimate universe. The alternative that the new physics hints to, that arises from an ontology based on the concept of event brings both poles together and allows a glimpse of a monist way to conceive the world that goes beyond the Cartesian dualism.

For the Cartesian mechanism, which conceives living beings as machines constructed to fulfill certain functions through the mere ensemble of parts, it is impossible to think that they were generated in a spontaneous evolutionary process. A mechanism is designed and built, and to the extent that it became possible to unravel the incredible complexity of living beings, the vision that they were designed by a superior intelligence impose itself. Let us recall that the vision of that time assumed implicitly that each species existed without change and required its own creation. As Jonas (2001) notes, modern science is born ignoring the problem of origin. It attempts to explain how things work, not how they came to be. The deism that predominated at the time, which assumed a preexisting God that created the world like a vast mechanism that once started continued working automatically complying with the laws of mechanics, protected the development of modern science in its infancy.

Since the laws of mechanics are completely deterministic and indifferent to the direction in which time advances, the past and future are in principle interchangeable and any notion of evolution appears impossible. Let us recall Laplace's statement (1814/2010): "We may regard the present state of the universe as the effect of its past and the cause of its future. An intellect which at a certain moment would know all forces that set nature in motion, and all positions of all items of which nature is composed, if this intellect were also vast enough to submit these data to analysis, it would embrace in a single formula the movements of the greatest bodies of the universe and those of the tiniest atom; for such an intellect nothing would be uncertain and the future just like the past would be present before its eyes." Only with the advent of thermodynamics a distinction between past and future arises, basically the one that results from the growth of entropy that predicts the passage from a more ordered to a more disordered sates. One of the first scientific treatments of origins appears with the nebular hypothesis of Kant and Laplace that says that the planets originated from a cloud of gas that orbited around the Sun. A merely mechanic process resulting from the movement of a system of gravitating particles produces through successive processes of condensation and cooling the objects of the Solar system that exist today. Kant's hypothesis has the implicit notion of thermodynamic phenomena in particular heat dissipation related with the increase of entropy.

19th century Darwinism seems to give an answer to the problem of the origin of living beings in terms compatible with the known laws without resorting to a creator or assuming teleological determinations reducing the problem to that of the origin of the first live being. The evolutionary process combining gradual variations and adaptation to a constantly changing environment seems to be able to generate superior life forms without assuming some type of pre-existence in the primitive organisms of the more advanced ones. Contrary to the preformationism that seemed implicit in the mechanist vision of life, the latter results from a process eminently contingent of exploration of possibilities only limited by the restrictions imposed by the laws of physics.

This process became more profound with the development of Freudian psychoanalysis and the attempts to explain mental phenomena in mechanist terms which, as we already mentioned, are at the basis of the nihilist conceptions developed at the end of the 19th century by eliminating any form of exceptionalism in relation with the human existence. Far from occupying the center of creation and of maintaining a privileged relationship with God who had created humans in its own image, humans were just animals with certain special abilities. In fact, Darwinian revolution not only undermined the traditional religious conception of mankind, it also reduced to the absurd the Cartesian "solution" to the mind/body problem and led to the philosophical enterprise of accounting for mental phenomena in mechanical terms, clearly impossible in its central objective. Indeed, by eliminating the need of any creative principle independent of matter to explain the emergence of animals superior species in purely physical terms, Darwinism leaves to this matter the task of explaining also the origin of the mind. In other words, of reducing the mental phenomena to purely mechanical causes.

Hans Jonas (2001) made a remarkable observation about this process: "The scientific advantage of dualism was, at its briefest, that the new mathematical ideal of natural knowledge was best served by, and indeed required, the clear-cut division between two realms which left science to deal with a pure res extensa, untained with the nonmathematical characters of being. That reality in toto was not of this one desirable kind had been realized by Galileo, whose doctrine of the mere subjectivity of the 'secondary qualities' (the expression is Locke's) initiated the extrusion of the undesirable features from physical reality. But subjects themselves are objective entities within reality, and the extrusion of features remained incomplete so long as their dumping-ground itself was part of the world to be described by natural science." Cartesian Dualism attempts to reduce to its minimal expression the inexplicable residue of the mental and at the same time isolate it from the external reality in order to make it completely describable in physical terms.

But the continuity discovered by the Darwinian evolution between mankind and the animals breaks the last attempt to isolate mankind as the only endowed with mental substance. Either one includes mental processes in the framework of a mechanist explanation of life, therefore completing the elimination of the sensitive sphere initiated by Galileo, or we understand that mental processes are not the exclusive purview of human beings.

As Jonas puts it: "For if it was no longer possible to regard his mind as discontinuous with prehuman biological history, then by the same token no excuse was left for denying mind, in proportionate degrees, to the closer or remoter ancestral forms, and hence to any level of animality..." But the mental phenomena, when extended to the rest of living beings must be emotional in the first place, fear or pain must be the most extended psychic qualities since they are directly related to survival and together with them the sensations associated with senses. It is precisely this type of simple mental states the ones that are most detached from cybernetic constructions. The process of revalue of life and the role of beings endowed of self-conscience like humans involves a radical change of perspective towards the traditional religious conceptions. We have not liberated ourselves yet of the primitive forms of conceiving a God, in particular the ones that resulted from the various ways of reacting to these traditional conceptions taking into account the first scientific discoveries. In spite of the multiple elements that point to new forms of humanism, a synthesis that summarizes the set of scientific discoveries that radically depart from the 19th century tradition that led to nihilism is still lacking.

Among these scientific discoveries it is worth highlighting the following. The revolution in physics that started with the notion of field and continues with the revision of the notions of space and time and the discovery that geometry is dynamical and culminates with quantum mechanics and the quantum theory of fields, transformed profoundly the notion of matter diminishing the breach between the physical, the biological and the psychic. The discovery of the molecular basis of live, of genomics and the epigenetic action of organisms and the understanding of self-organizing phenomena that show that we live in a Universe with special capabilities to harbor and develop vital phenomena. The progressive understanding of emergent phenomena and downward causation that appear omnipresent in development of life and conscience. The spectacular advances of cosmology based on the detailed study of the behavior of the electromagnetic radiation remnant from the Big Bang and the discovery of dark energy responsible for the accelerated expansion of the universe. These discoveries and others that we will analyze in the following chapters were accompanied, in the last few decades, by the growing conviction that we live in a universe with laws particularly suited to harbor life. The centrality of live in our Universe and its implications is what we need to understand.

Chapter 12

Ontology of events and consciousness

12.1 Back to consciousness

Physics, and science in general, is the result of a drive towards liberating us from our senses, which started with Galileo, as we have discussed. The very effort for having realist interpretations of the quantum theory has as a goal a description where the role of observers is irrelevant. All descriptions where the observers, their measurements and experiences play a central role have been laboriously eliminated from physics. Realist interpretations, like the ones discussed here, can be seen as the last step in that direction.

Let us recall that Galileo considered that the basic concepts to understand the idea of matter are those of number, space and time. He claimed that "the universe's book is written in the language of mathematics." He was the most important modern figure to revive a distinction already present in Greek atomists between (what later would be called) the primary intrinsic qualities of the object, and the secondary ones associated to the sensations of color, smell, taste or sound. These designate the changing states of our mind. He says "let us suppress mentally the living beings and their organs and such qualities will disappear from the world."

The development of physics starting from classical mechanics up to today implied a process of enrichment of the basic ontological objects. For instance, compare the idea of object of classical mechanics with the one we here considering characterized by systems that are in states that manifest themselves in events. There is a parallel progress in the mathematical framework, which evolves from a description in terms of extended objects given by positions and velocities in a Euclidean space, to a description in terms of states and properties in an infinitely richer Hilbert space. As we will see, this could lead to understand consciousness in physical terms.

The limits within which physics operates are not surprising. After all, it was developed to describe universal aspects of reality that can be objectively described by strict mathematical relation. It therefore should not appear as strange that the strategy so developed cannot be applied successfully to mental processes. As Nagel (1994) observed, the methodology of physics "can be used on the body, including its central nervous system, and on the relation of neural activity to observable behavioral functioning, because they are all aspects of objective physical reality. But for the subjective qualities of experience themselves, we need a different form of understanding. We cannot hope to understand them completely, as an aspect of complete physical reality, because the concept of physical reality depends on excluding them from what has to be understood." Is the physicalist position held in this book therefore untenable? How can we cast aside the subjective aspect of our experience? How does one make compatible the naturalist description of the universe including our brains with their mental processes, in particular consciousness? To understand these difficulties is without a doubt one of the biggest scientific challenges that we must face.

12.2   What do we mean by consciousness?

The word consciousness has many meanings. In some occasions we refer to our ability to know our mental states and report about them. In others we simply refer to our ability to notice that something happens or to learn something about it. These usages of the word do not refer to the most problematic aspect of the term: that in consciousness there is also an internal aspect, that refers to a cognitive agent. This is already pointed out by Socrates, in his dialogue with Theaetetus when answering a question by the latter about "what is color?". He says that the color is "peculiar of each percipient" and he asks the young student "are you quite certain that the several colors appear to a dog or any animal as they appear to you". Theaetetus responded negatively to which Socrates adds "or that anything appears the same to you as to another man?" This eminently private and subjective aspect is part of the conscious experience. It includes the vivid sensations of color of a morning in the countryside as well as its subtle scents.

The key point behind the conscious experiences, as Socrates' question suggests, is that "there is something it is like to be in that organism" as is asked by Nagel (1974) in his famous article "What it is like to be a bat." There are unique and nontransferable experiences that each of us have, there are qualitative feelings, also known as phenomenal qualities or simply qualia (Chalmers 1996). This property of the mental entities, although not the only one involved in the conscious process, has been the hardest one to explain. In spite of this last point, it surely should not be possible to explain qualia in isolation, because the phenomenal is always accompanied by the psychological.

This latter aspect of the mind is at the root of its behavior. Only if what is felt (phenomenal) and what the mind does (psychological) is included in the description, we will be in position to account for mental phenomena. Whereas what is felt is private, what the mind does about it may be public as it relates to behaviors. In the end, as Chalmers observes, there do not exist other phenomena that we must account for other than "those we have third-person access to, like atoms or chairs and those we have first person access to like pains, colors or itches. The former are external and inter-subjective, that is, any observer in appropriate conditions can access them. The latter, on the other hand, are internal and only accessible to the individual that experiments them."

John Searle (2007) adds another fundamental property of our conscious states: they are experienced by us as part of a single unified conscious field. "Your conscious states at any moment are parts of a single big conscious state. The visual experience of the tree, the tactile experience of the desktop under my hand, and the sight of the moon outside my window are parts of a single total conscious experience. But other entities in the world are not like that. The tree, the desk, and the moon are not in that way parts of a single total big object... For a state to be qualitative and subjective implies that it is part of a unified field of qualitative subjectivity, even if it is the only thing in the field. If you try to imagine your present conscious field broken into seven parts you will find yourself imagining not one conscious field in seven pieces, but rather seven separate conscious fields."

Any attempts to explain consciousness in naturalist terms, even preliminarily, must also take into account that: 1) Conscious contents are ineliminable, we have direct evidence of our conscious states. A conscious content does not stop being even if we can prove that it is a mere appearance without no objective reality. They are ineliminable and incorrigible. Incorrigibility of the mind is the view that "if you believe you are in a mental state, the belief cannot be false." (Priest 1992). 2) Admitting a naturalist conception for them, our conscious contents must be associated to states of our nervous system. If we do not discard the possibility of conscious states in other situations, the general hypothesis must be that every conscious state has a physical counterpart. 3) Our conscious states have causal efficiency. Many of our acts are explainable in terms of our mental processes. There exist correlations between our mental and neural processes and in some cases some form of top down causation. For instance the one described when we treated the emergence of complex quantum systems, that accounts for the efficacy of the mind to induce changes in physical behaviors.

12.3   The problem of consciousness and quantum theory

In the previous chapter we have introduced the principles of a quantum mechanical ontology. It is an ontology at the moment incomplete and too tied into the axiomatics of quantum mechanics. It is based on the objective vision of quantum reality that we have developed in some detail in the second part of the book. It adopts a realist position about the theoretical entities based on an interpretation of quantum mechanics and therefore up to now it only includes such phenomena for which we have "a third person access." Its fundamental elements are systems, states and events.

Events have been identified as the building blocks of apparent reality and everything that surrounds us can be considered as composed of elementary events. Events are therefore physical entities —included in the framework—, totally independent of any measurement. They happen and connect with each other causally. Macroscopic phenomena are associated with bundles of events organized in space and time. Events characterize the apparent side of reality, that is, what is accessible to our experience. On the other hand,quantum individual things, which we associated to the systems in given dispositional states, are in general holistic and non-local. Whereas events describe the actual appearance of objects, the individual things are associated to their potential behavior with respect to other systems, not just with the events they produce.

Since there is a growing consensus that we live in a purely quantum universe, the adoption of a quantum ontology appears rather inevitable. This is independent of some claims of macroscopic quantum coherence in the neurological phenomena of interest (Penrose and Hameroff 1995). Indeed, in a quantum universe the phenomena described by classical physics are particular cases of quantum processes about which we can predict its behaviors, and therefore, there should have the same ontological character. The behaviors that take place in a photographic plate, a table, or the brain, should be considered of the same fundamental nature, and therefore described by systems in determined states that produce events.

It has been noted some time ago that an ontology of events allows a promising approach to the problem of consciousness. The position known as neutral monism holds basically that events can be associated both to physical and phenomenal aspects. It is a particular case of the double aspect theories in which one class of entities has both physical or third person manifestations and phenomenal or first person manifestations. That is, each event is something for itself and something for other events. In its origins neutral monism can be attributed to Mach (1886), but it has illustrious predecessors that held, like Spinoza (1677/2005) that the physical and the mental are two aspects of the same substance. Although our position differs significantly from neutral monism, it is important to recognize that it stems from an elaboration on its ideas.

Throughout his life Bertrand Russell adopted different positions close to neutral monism. For instance in 1927 (Russell 1927/2007) he said "Matter in a given place are all the events that are there..." and he goes on to say in the same book that such vision of matter implies that we "do not have to deal anymore with what used to be mysterious about the causal theory of perception: how a series of waves of light or sound produce an event apparently totally different from them in its character." That is, they produce a sensation. As we have seen when we studied quantum ontology, matter does not reduce to pure events. In similar fashion, mental contents do not reduce to mere sensations, as we will analyze later in some detail. But nevertheless, sensations play an essential role for their understanding. In the next passage Russell is even more emphatic in his unified vision between the mental and the physical: "I think... that an ultimate scientific account of what goes on in the world, if it were ascertainable, would resemble psychology rather than physics... such an account would not be content to speak, even formally, as though matter,which is a logical fiction, were the ultimate reality. I think that, if our scientific knowledge were adequate to the task, which it neither is nor is likely to become, it would... state the causal laws of the world in terms of... particulars, not in terms of matter. Causal laws so stated would, I believe, be applicable to psychology and physics equally; the science in which they were stated would succeed in achieving what metaphysics has vainly attempted, namely a unified account of what really happens, wholly true even if not the whole of truth, and free from all convenient fictions or unwarrantable assumptions of metaphysical entities."

A first point to notice therefore, just as a preliminary verification stemming from observations like those of neutral monists, is that in a quantum ontology, conscious phenomena as sensations could be associated to events in our brain to which we have a first-person access. In other words, whereas we only have an indirect access to the events of the systems we study physically, that is we know of them ultimately by their effects on other objects and eventually on us, we have direct access to the events in our brain as long as we are aware of them. In that sense we can say that we only have idea of how things are in themselves through our conscience. If the events have a phenomenal aspect that manifests itself when we have a first person access, does something similar occur with the states? Russell only refers to events since following a long empiricist tradition, he considers the world as composed entirely by them. In his book on the mind (Russell 1921/2011) he characterizes neutral monism by identifying sensations with a particular class of events: those that occur in our brain. With that aim he attempts to show that mental phenomena reduce to sensations and images, the latter being less vivid sensations that arise in our brain as reproductions of the original ones. Therefore he establishes a clear asymmetry that he seems to notice, at least momentarily.

Indeed, in spite of observing at a certain point that "Since the mental world and the physical world interact, there would be a boundary between the two: there would be events with physical causes and men-

tal effects, while there would be others which would have mental causes
and physical effects. Those that have physical causes and mental effects
we should define as "sensations". Those that have mental causes and
physical effects might perhaps be identified with what we call voluntary
movements; but they do not concern us at present". Russell in fact
postpones the discussion of the will until the last paragraph of the next
to last chapter of the book and, concludes after a brief analysis, that
the volitions do not require the addition of any extra ingredient to the
analysis of the mind.

In a moment we shall see that what we describe as emotions and
volitional acts could be related with we have described up to know as
dispositional states. Precisely this volitive aspect or more generally the
intentionality is what could be associated to states. Searle describes
intentionality in simple words as follows: "My states of thirst, hunger,
and visual perception are all directed at something and so they fit the
label of being intentional in this sense. Undirected feelings of well being
or anxiety are not intentional."

As we have seen when we discussed the ontology of events, states
cannot be left aside in a complete description of the world. The fundamental physical laws are not formulated in terms of chains of events.
Recall that in quantum mechanics the Schrödinger equation is the main
tool for describing the evolution of a system and refers to states and
not to events. Objects like an atom are primarily dispositions to behave
in a certain way. That essential character, together with other characteristics of the states we have already mentioned, all suggest that the
states can have the first person aspect that one usually associates with
the concept of intentionality. They characterize our disposition to act
on other systems.

Notice the similarities existent between the characteristics of mental
phenomena: intentional, incorrigible and private, and those of the states.
The latter are dispositional and inaccessible by isolated measurements,
that is, they are private, and they may have an internal aspect like
the one we assume here. As we have emphasized throughout the book,
states characterize the disposition of the system to act producing certain
effects on other systems. To characterize the disposition of a state to
act on any other system is to give its most complete description. States
are private in the sense that one cannot determine in which state is
a system by measuring it. One needs an ensemble of identical states
and measurements on each member of the ensemble in order to have a

complete determination of the states. As it does not make sense to have an ensemble of identical mental states, they are inaccessible to external observers.

If things are as we are laying them out here, both events and states would have a phenomenal aspect and a third person aspect that we are aware of. In the case of states, they would also have intentionality. One may therefore think that in all physical systems there could be a form of "proto-consciousness". However, in classical systems, where the loss of coherence is very efficient, the intentional or dispositional aspect is not present and the system is deterministic. Chalmers claims that awareness with phenomenal experience is enough to have some form of consciousness. By being aware he understands that the system have access to some information and can use it to control its behavior. It does not appear that the mere possession of information nor the possible existence of phenomenal experiences are enough to have consciousness. Let us consider an extreme example: there exist very simple control devices, like electronic sensors, thermostats, pressure regulators, that are systems that process information. To an input event they can associate an output behavior, for instance turning on or off an air conditioning unit. This does not differ from a classic deterministic process. The only difference is that we are interested in describing it in terms of information that is registered and correlated with an output behavior. As long as the device operates correctly we have a process where information is preserved. Even admitting the existence of an internal "proto-phenomenal experience" associated with each event this appears poorly related with what occurs in a human brain or even perhaps in the nervous system of an animal. Criticism about this view generally relies on naive arguments that view the internal aspect of events in terms of familiar mental properties: how could an atom be aware or feel pain. This is not the kind of phenomenal experience that might be associated to the phenomenal aspect of a quantum event like the appearance of a dot due to the arrival of electrons in a photographic plate.

One could think that this is the pan-psychist point of view, the one that claims that everything has a mind. This point of view appears incorrect. Proto-phenomenal properties in isolated events have very little to do with mental phenomena. Even accepting that there exist phenomenal aspects in simple systems, it would be a serious mistake to claim that complex phenomena would arise from mere collections of such elementary phenomena. It would appear that, in what respects to the human conscience, there is a new level of holistic integration —as we have already stressed when we mentioned the unity of the conscious experience— like the one we recognized in the emergent biological and chemical processes. In every instant a countless number of sensations converge in our conscious sphere, that one should think of as associated to the instantiation of properties of very complex events. Perhaps the multiplicity of mental qualities of which we are aware of in a given instant implies the existence of a single event, enormously rich in properties and phenomenal contents, for every instant of consciousness. If this is so the conscious states could potentially be associated to highly entangled states in quantum systems present in our brain.

A point that is not clear is if the very complex systems in our brain can keep a holistic behavior without suffering an instantaneous loss of the coherence needed for the required quantum behavior. Up to now, there is not any evidence of processes in our brain that can give rise to quantum phenomena with large scales of coherence. Several suggestions of possible mechanisms have been done Penrose and Hameroff (1995) Mavromatos (2011), however at the moment these effects have not been conclusively confirmed The increasing evidence of quantum phenomena in biological systems, (e.g. Al-Khalili and McFadden 2016) suggests that they could be relevant to the functioning of our brain. For instance, recently different studies show that quantum coherence could be relevant in photosynthesis (Collini et al. 2010) and the development of cancerous tumors (Plankar et al. 2011).

The claim that quantum physics plays a central role in consciousness is obviously still contentious and significantly more research is needed to confirm a particular standpoint. Our intention here was just to point out that the special qualities of quantum physics could perhaps help solve this conundrum about the unity of the consciousness states. *But even if large scales of coherence were not present in our brain, the fundamental physical description of the brain as of any physical object is quantum mechanical and therefore it should be analyzed with the concepts of the quantum ontology as we have proposed.*

12.4 The mind-body problem and the quantum ontology

Thomas Nagel characterizes the mind body problem in the following terms: "The mind-body problem exists because we naturally want to include the mental life of conscious organisms in a comprehensive scientific understanding of the world. On the one hand it seems obvious that everything that happens in the mind depends on, or is, something that happens in the brain. On the other hand the defining features of mental states and events, features like their intentionality, their subjectivity and their conscious experiential quality, seem not to be comprehensible simply in terms of the physical operation of the organism. This is not just because we have not yet accumulated enough empirical information: the problem is theoretical. We cannot at present imagine an explanation of color perception, for example, which would do for that phenomenon what chemistry has done for combustion–an explanation which would tell us in physical terms, and without residue, what the experience of color perception is. Philosophical analyses of the distinguishing features of the mental that are designed to get us over this hurdle generally involve implausible forms of reductionism, behaviorist in inspiration. The question is whether there is another way of bringing mental phenomena into a unified conception of objective reality, without relying on a narrow standard of objectivity which excludes everything that makes them interesting."

We would like to explore the possibility that the quantum ontology we have introduced could contribute to the understanding of the mind-body problem in physical terms. To this aim it is good to start with a discussion of the concept of object and substance in philosophy and how they relate to the basic concepts of the ontology we introduced.

12.4.1 Objects and substances

According to the influential conception of Locke, objects are entities which possess or "bear" properties. Reciprocally, properties are borne by objects. In that sense, objects are entities which bear properties but which are not "borne" by anything. Objects are therefore considered as the substratum of properties and therefore identified with substances. According to Locke substance in general is a bearer of properties and a cause of properties. Besides this characteristic one cannot say anything about substances. For, instance, Locke himself said that substances were "Something, I know not what". This identification of the substance as a mere substrate without qualities was an easy target for objections. Indeed, conceived in that way, they have an absolutely non-empirical character, since every time we say we observe a substance we cannot make reference to anything else but its properties. For that reason Hume abandoned the concept of substance and went on to claim that the objects are nothing more than bundles of properties that change gradually in time and to which the mind assigns certain unity.

The quantum theory adds a completely new point of view concerning the nature of objects, in particular it sheds new light on many of the ambiguities that accompany the concept of substance since the times of Locke. In the quantum ontology the individual objects, that is the systems in given states, do not have properties like Locke's substances, they only have dispositions to produce events and their corresponding properties with certain probability. But contrary to the notion of substance as an undifferentiated substrate, here each individual object has characteristics that identify it as a well defined system with a well defined dispositional state. This elucidates the notion of substance, becoming concrete and empirically analyzable. The production of events by an individual thing always depends on the environment that surrounds it. As we have seen when we discussed the axioms of quantum mechanics, events take place in a system that includes the individual and the environment. Every time an event takes place the individual changes its state and its future dispositions. In a macroscopic object, like a chunk of iron, the interaction with the environment, even in vacuum, is very effective and the production of events occurs in practically continuous fashion. For instance Tegmark (1993) computed that even a speck of dust in outer space close to the Solar system, would produce thousands of events per second. The effect grows in proportion to the surface area of the object.

The definition of object as a bundle of properties, which is the most popular one among empiricists and analytic philosophers, leaves totally aside the underlying dynamical aspect of any physical phenomenon, which the laws of quantum mechanics describe in terms of an evolution of states obeying the Schrödinger equation rather than a succession of events. Many philosophical problems become unsolvable if one abandons the notion of state. A problem we have already studied and seen how is resolved in terms of states is that of emergence.

12.4.2   Properties and events

When empiricist philosophers like Hume think of an object, they consider the totality of its properties: color, shape, texture, etc. They describe them as a thing and assign them an identity. Although the totality of properties present give the illusion that there exists something more, they only have evidence of the properties. Studying the fundamental objects of quantum physics we were able to identify in an unambiguous and empirically verifiable way that there exists a substrate: the object. The second element we found was that the properties instantiate via the events. Every time that an event takes place, like the appearance of a dot on a photographic plate, the event is accompanied by a set of properties that as we saw in previous chapters, the formalism characterizes precisely.

The appearance of the event is the cause and the mechanisms for new properties to be instantiated. This effectiveness of the event as a process through which reality becomes apparent and properties manifest themselves, is the basis of the world of by classical physics, that is, a world of well defined properties. The vision of quantum physics rids us of a second philosophical problem associated to the conception of an object as a bundle of properties. This problem was first recognized by Hume. If the objects are no more than bundles of properties the only laws that can be recognized are those associated to constant successions of events. John Stuart Mill (Stumpf 2007) says about this: "We have no knowledge of anything but phenomena; and our knowledge of phenomena is relative not absolute. We know not the essence, not the real mode of production, of any fact, but only its relations to other facts in the way of succession or similitude. These relations are constant; that is, always the same in the same circumstances. The constant resemblances which link phenomena together, and the constant sequences which unite them as antecedent and consequent, are termed their laws. The laws of phenomena are all we know respecting them. Their essential nature and their ultimate causes, ...are unknown and inscrutable to us."

But this is not the way physics proceeds, in spite of being an empirical science. The events are not directly related with other events. They are related indirectly through the states of the systems that produce them. The laws of dynamics, like Schrödinger's equation, refer to states and not to events. As a consequence, the laws are statements about tendencies or dispositions of objects to produce events, they are not statements about how events follow one another. Therefore, it is not surprising a law, in order to be effective, should basically establish connections that involve all events in the universe that have occurred in the past. The language of states allows us to describe the system in terms of the present available information. For instance, in order to know the state of an atom at a given instant, one needs to know not only the previous events induced by the atom, but its interaction with other systems. This interaction is described via the Schrödinger equation in terms of interaction forces which also depend on the state of the relevant portion of the environment. In terms the state of the system condenses all the relevant information to describe the future behavior of the system. In ontological terms the state represents the disposition of the system to behave in one way or another according to the circumstances it finds itself in.

12.4.3 The mind-body problem

The mind-body problem deals with the relation of mental and neural events. We have anticipated the idea that the states and events can have an internal aspect. We would like to be more specific concerning this statement. The mental events have subjective and phenomenal aspects. On the other hand, for events in the brain we agree that although our understanding of them evolves, their regularity will be describable in physical terms that surely will involve biological and electrochemical phenomena and ultimately quantum mechanical phenomena.

We have also suggested that certain events that occur in our brains are responsible for our mental states in a relationship so intimate that it only admits a distinction in the way we approach those events. Either from outside by a neurophysiologist or from inside by the subject being studied. In what sense can we therefore claim that the brain and mind events are identical? The simplest hypothesis is that of coincidence. Certain mind events that occur at a given instant can be associated with brain events that occur simultaneously. We cannot clearly say that both events are identical in their properties, since in one case they will be expressible in terms of physical magnitudes and in the other by perceptions, memories, decisions, etc. The two ways of accessing events in our brain give rise to different properties. The mathematical description of properties given by quantum mechanics in terms of projectors refer to their objective or third person character.

The fact that the objects in this ontology emerge from the production of events by the individuals frees us from the idea of object as a bundle of properties and allows a theory where the objects so conceived have a double aspect. This theory is an evolved form of parallelism as the one first proposed by Spinoza. The concept of object in terms of states that produce events allows to address an even more disquieting point of the mind-body problem: it is very difficult to imagine that mental facts, as we ordinarily conceive them, do not intervene in the explanation of our actions. Most humans have the conviction that "There are no strict deterministic laws on the basis of which mental events can be predicted and explained" (Davidson 1970).

If we circumscribed ourselves to a classical physical description, this would immediately contradict a very extended intuition. Brain events, just like any other event in physical nature, are subject to laws. In a deterministic world that would imply that mind aspects are also subject to deterministic laws. The classical description of the world is complete in the sense that given the initial state of the system all its properties at a future instant are determined by the initial state and the laws of classical physics. Also, the properties of a system are supervenient on the properties of its components. In other words, a classical world is closed causally. Even though in a regularist scheme one assumes that the laws of physics do not exhaust reality and only describe it as long as it follows regular behaviors, it is not conceivable to have mind events that are non deterministic consistent with the totally deterministic occurrence of brain events.

In a quantum world with probabilistic laws there is room for a non deterministic connection among mental events. There is no causal closure in the physical world as in the classical case because future events are not uniquely determined. This would be true even if it were possible to have a complete knowledge of the state present in the brain, something that the quantum theory does not allow anyway. This observation is at the core of the proposal that we will outline in this chapter

Our approach, that we propose to call regularist naturalism, can be summarized in three hypotheses:

1) *Double aspect of the fundamental entities.* The fundamental entities of the quantum ontology that have been identified up to now, events and individuals, have a double aspect. A first person aspect that manifest itself as sensations, emotions or thoughts and a third person aspect that manifests itself in the neural states and events. Not only the

events must have an internal aspect: the sensations. The dispositions of the states may also have an internal aspect manifested in emotion or volition. Both aspects manifest themselves in properties associated to the same entity of phenomenal or physical character. The correlation is fundamentally of ontic character, that is, derived from the nature of things. To establish that the correlations are "nomic" that is that are described by explicit correlation laws as philosopher like like Honderich claim appears unsustainable when one of the aspects the phenomenal one— does not allow a third person access. This therefore makes objectively incomparable the phenomenal properties that occur in each case. It is not possible, for example, to compare the "yellow" that two people see when they look at a fried egg. As long as laws define objective empirical regularities, they cannot be applied to the phenomenal realm. If we understand that each fundamental entity, events and states, have a first and third person aspect, we will be able to speak about psychoneural pairs but in the end we will be talking about the same event or thing from two different perspectives.

2) *Downward causation.* What explains the efficacy of desires or reasons in the determination of our actions. This question is intimately related with the issue of emergence and top down causation It is clear that emergence should imply downward causation in order to do some serious causal work affecting the evolution of the physicochemical components that contribute to a consciousness state. We have already mentioned that strong emergence leading to higher level phenomena having causal powers on the lower level phenomena are impossible within the context of classical physics but are naturally present in quantum mechanical systems like atoms and molecules. A similar mechanism should be required between the mental processes and their neuronal basis which once more suggests that quantum mechanics should play a fundamental role in the understanding of mental causation. The downward causation of mental states is explainable in physical terms in the same way that entangled states of composite systems can be described in physical terms. However, the downward causation, as we have already stressed, implies new properties and new causal power appear at the upper levels which are ontologically non reducible.

3) *Regularism.* Physical laws in quantum mechanics are probabilistic. Even if one admits that the evolution of the neural states one can use Schrödinger's equation which is deterministic, the production of events in any of the interpretations of quantum mechanics discussed

before are random, and therefore not ruled by any law The choice of an event is only limited probabilistically. This opens the possibility of introducing a concept associated with brain states that precedes the production of events and that is not nomologically regulated, which we call free choices. Among the interpretations of quantum mechanics that admits an ontology in terms of events there are some —like the many worlds interpretation— that do not admit free choices. Others like the Montevideo Interpretation or the Modal interpretation allow to speak of free choices. Recently, Mauro Dorato (2015) has suggested that many popular interpretations admit an ontology of events, among them some antirealistic views about the wave function like Bohr's version of the Copenhagen interpretation that also admit speaking of free choices.

The three hypothesis sketched above about the nature of the mind add a richer vision than those of the traditional mechanistic positions and may allow to overcome the epiphenomenalist challenge that reduces consciousness to a "mere subproduct of the bodily behavior, lacking in all capacity to modify that behavior, the same way that the sound of the steam siren accompanies the working of a steam engine... [although] it lacks any influence on its machinery" (Huxley 1893/2001). The psychological world is not a mere epiphenomenon of the physical world, it transcends it in total analogy to how the "things in themselves" transcend their physical description. The interiority of the subject becomes effective when the nomological deductive description of physics does not give complete information about its behavior.

The incompleteness of the physical objective account has been repeatedly noticed by thinkers on the problems of consciousness, for instance Tye (2007) says: "no fully objective mechanism could close the explanatory gap between the objective and the subjective. No matter how deeply we probe into the physical structure of neurons and the chemical transactions which occur when they fire,..., we still seem to be left with something that cries out for a further explanation, namely why and how this collection of neural and chemical changes produces *that* subjective feeling or any subjective feeling at all" We consider that this incompleteness may not only refer to the phenomenal character of consciousness but also to the capacity of the subject to chose among possible events that are only limited probabilistically. We will come back to this point in the forthcoming chapters.

Chapter 13

Cosmology: the genesis of a bio-friendly universe

13.1 Synopsis of the evolution of the Universe

Our ancestors gazed at the skies since time immemorial, seeing the Sun, the Moon and the stars moving around the Earth at unknown distances. Only recently have astronomers been able to find the size of the Universe. Our Sun, together with hundreds of billions of Suns, constitutes our galaxy, the Milky Way. It is only one of the hundreds of billions of galaxies that we can see through our telescopes. In terms of distances, light takes only one second to go from the Earth to the Moon and eight minutes to reach us from the Sun. The Milky Way is vastly bigger than our Solar System, a light signal would take some 100,000 years to go from one end to the other of the galactic disk. The farthest visible galaxies are at about 10 billion light years.

In short, the Universe is very big and matter is distributed in a roughly uniform fashion. Certain inhomogeneities with clusters of galaxies and almost void regions can be seen that are expanding and therefore becoming bigger.

At very large scales the distribution of galaxies is very similar in the various regions of the Universe. Hubble and Slipher discovered in 1929 that light arriving from far away galaxies redshifted, becoming more

209

redshifted the farther the galaxy it originated in. We know that an object that moves away from us redshifts and one that moves towards us blueshifts. Therefore we conclude that the farthest galaxies move apart from each other. The farther the distance the bigger the separation speed.

As the Universe expands, in the past it was much hotter and denser, as would happen to a gas that is compressed. Going enough into the past the Universe was so hot that atoms could not exist, they would break up into individual protons, neutrons and electrons. At that stage the Universe was a very hot plasma composed of particles.

The dynamics of the Universe is governed by general relativity. We know matter deforms space-time, bending it near massive objects. Gravitational forces are explained in terms of masses that move in a curved space. General relativity can be applied to the Universe as a whole and allows us to conclude that the expansion started with a great explosion: the Big Bang. Somewhat less than 13,800 million years the Universe was infinitely dense and hot. When physical magnitudes become infinite it is a sign that the theory being used has stopped being valid. Indeed, in order to describe the first instants of the Universe it would be necessary to use a theory that combines quantum mechanics and general relativity. Of this initial period that lasted an infinitesimal fraction of a second we cannot say much at present, except that the Universe was extraordinarily dense and energetic and, according to our current understanding, all the forces of nature were unified.

After a small fraction of a second a phenomenon starts to take place — on which we will elaborate soon— known as *inflation*. At that stage the size of the Universe grows exponentially. At the end of the inflationary process the Universe starts to form the types of matter we know today. First the fundamental components of protons and neutrons form, which constitute a quark-gluon plasma at very high temperature. As the temperature lowers the components of atomic nuclei form: protons and neutrons. A minute after the Big Bang the protons and neutrons form bound systems and give rise to the nuclei of the isotopes of hydrogen: deuterium and tritium and the light elements, fundamentally helium and some lithium. 370,000 years later electrons combine with the nuclei to form atoms. At that point matter becomes electrically neutral and the *universe became transparent*. From there on light could travel without being immediately absorbed by matter, as it still is today.

We have mentioned that the Universe was very dense and hot near

Figure 13.1: The Rosetta stone of cosmology. The Universe in its infancy 370,000 years after the Big Bang.

its beginning. We know that hot matter emits radiation, like does the Sun or a very hot piece of iron. It therefore was expected that this radiation stemming from the Universe at the time atoms were formed would eventually reach us. In 1967 two radio engineers in Bell labs, Arno Penzias and Robert Wilson, found radiation incoming from space that became the strongest confirmation of the Big Bang. The radiation is distributed almost homogeneously in the sky and corresponds to a source that emitted at 2.725 degrees Kelvin. This is due to the cooling that the radiation emitted 370,000 years after the Big Bang suffered as the Universe expanded to its current size. Initially its temperature was several thousands of degrees. That radiation, known as the "cosmic background radiation" allows us to see how the universe was in its beginnings 370,000 years after the Big Bang. When observed with an optical telescope the space between stars and galaxies appears to be completely dark. However, a radio telescope shows a faint background glow, almost isotropic, that is not associated with any star, galaxy, or other object. The map given in the figure es like a 360 degree photograph of the oldest light in our Universe, emitted when the Universe was 370,000 years old. It shows small temperature fluctuations that correspond to regions of slightly different densities, representing the seeds of all the present structure: the stars and galaxies. The red spots represent slightly hotter regions of the sky and the blue colder regions.

13.2   The dark universe

Today we know that in addition to the ordinary matter —protons, neutrons, electrons, light, neutrinos...— the Universe contains two other
forms of matter. The latter can only be detected through their gravitational effects and are known as *dark matter* and *dark energy*. This has
been a surprising discovery of the last two decades, although the existence of invisible gravitating matter was first theorized 70 years ago by
Fritz Zwicky. He had observed that the dynamics of clusters of galaxies
could not be explained in terms of the effects of visible matter, which
suggested the existence of some type of exotic matter not visible by our
telescopes. Today we know that the dynamics of stars in galaxies also
cannot be explained without assuming the existence of forms of matter not detected yet. Several hypotheses about the origin of this form
of matter have been put forward. Later, direct effects of dark matter
on light were observed: light bends while traversing dark matter as if
it were traversing a lens. Dark matter interacts very little with ordinary matter, it only interacts gravitationally and probably with weak
interactions with other forms of matter.

The recent detection of gravitational waves from very massive col-
liding black holes makes it more plausible that a significant part of that
matter is due to the existence of black holes. Other hypotheses have
been put forward, for instance the existence of WIMPs, weakly interactive massive particles like neutrinos. Neutrinos themselves have masses
and abundances that are not enough to account for the required dark
matter.

Today we know that dark matter played a key role in the formation
of the Universe providing most of the gravitational force needed for the
formation of galaxies. We know today that less than 5% of the mass of
the Universe is ordinary matter and the rest are dark forms of matter.
We also know that the kinds of dark matter we mentioned up to now
are only one third of the total dark mass. The rest comes from a form
of matter even more exotic with defining effects on the evolution of the
Universe.

In the mid 1990's it was observed that the rate of expansion of the
Universe accelerates, as if in addition to gravity there existed a repul-

sive force. The Universe expands today faster than in the past. The initial hypothesis of the Big Bang was that after the great explosion the rate of expansion would diminish due to gravitational attraction. We now know that an antigravitational force opposes gravity, vanquishing it and accelerating the expansion. Almost a century ago Einstein had proposed a force of this kind as part of the equation that the proposed to describe general relativity. He called it the *cosmological constant*. Einstein later regretted including that term that some say he called his "greatest blunder". The observations of the last decades suggests Einstein was not entirely wrong.

The consensus of many cosmologists today is that the final fate of the Universe depends on its global shape and how much dark energy it contains. The Universe seems to contain just the needed amount of dark energy in order to be spatially flat. We again encounter a form of matter that manifests itself through its gravitational effects, in this case repulsive, which has been called *dark energy*. In any event, the contribution of dark energy to the total mass of the universe exceeds the sum of normal matter and dark matter. If the density of dark energy remains constant in the expanding universe as it appears to be, it will become the dominant form of energy in the future Universe. If this were the case the galaxies would recede from each other at a faster and faster rate and the light from them would become redder until their apparent speed of recess seems to exceed that of light in which case there would not be visible anymore.

13.3 Inflation

We have noted that as we travel into the past the Universe becomes hotter and the reactions that take place become more and more energetic. Current particle accelerators like the Large Hadron Collider (LHC), which allowed to discover the Higgs boson, allow to reach energies similar to those existing at a trillionth of a second after the Big Bang. In the first instants of the Universe changes become faster and faster, for instance the temperature or the energy at the time of the Big Bang were infinite but in a trillionth of a second it gets down to what our current accelerators can produce. It is therefore not surprising that understanding what happened in that initial epoch is very important not only because the most exotic and difficult to understand phenomena took place then but because the dominant properties of the Universe were defined there. In that initial epoch is when inflation must have taken place.

The expansion and cooling governed by general relativity allow to explain, as we have seen, many things. However, towards the 1970's there were left some important problems that remained unexplained. To begin with, although near massive objects space-time is curved, the Universe at large scales appears perfectly spatially flat. An explanation was lacking as to why the Universe contained the exact amount of matter necessary for it to be spatially flat. General relativity predicts three possible behaviors depending on the amount of matter in the Universe. If there is a lot of matter the spatial geometry is spherical, light rays from distant sources form angles of more than 180 degrees. If there is too little matter the geometry is hyperbolic and the angles add up to less than 180 degrees. For the geometry to be flat the amount of matter must have a precise intermediate value. What made the amount of matter just the correct one for the Universe to be flat?

On the other hand, the Universe appears extremely uniform no matter in which direction we look. For example, the cosmic microwave background radiation has a mean temperature of 2.725 degrees Kelvin with fluctuations of a few ten-thousandths of a degree. Ha can the Universe behave like a system that is in equilibrium with the same temperature in all directions and with fluctuations so small? In a second we will explain why this is so difficult to understand. Before that we observe that these apparently inexplicable properties are crucial for the Universe being adequate to harbor life. If the Universe expanded too fast as it will if it had less density of matter the gases of the primordial cosmos would not have had enough time to form galaxies. If it had more density of matter the gravitational attraction would have been bigger and the Universe would recollapse before enough time for life to establish itself. Our Universe has the exact amount of matter for having a large number of galaxies for a long enough period of time to harbor life.

The difficulty to explain the homogeneity of space is explained by Paul Davies (2006) in the following way: "Imagine two regions of space on opposites sides of the sky A and B. Each 10 billion light-years from Earth. They look very much the same, containing similar distributions of galaxies with similar red shifts. Located poles apart, these cosmic regions are seen by us today to be separated by about 20 billion light-years from each other. Since the universe is less than 14 billion years old, light cannot have traveled from one region to other. An inhabitant in region A would not been able to see region B or know of its existence... The horizon around A will not yet have extended as far as B. Regions A and B evidently cannot know about each other. This means that they are what is termed causally independent, because no object or force can travel faster than light, so no physical influence can have linked these regions." How can we justify that two regions that never had a chance to interact had reached an apparent equilibrium and acquired absolutely similar properties? When the cosmic background radiation was formed one can see that there were millions of causally independent regions, therefore why does radiation stemming from those regions appear so uniform?

Alan Guth showed that these properties can be explained in a simple way if one assumes that the Universe underwent a stage of very rapid expansion called *inflation*. In a small fraction of a second the Universe's size was multiplied by a large factor: the end size was trillions of trillions of times the original one. Guth noted that although the Universe could have started with an arbitrary density and a random distribution of temperature, a spectacular growth in size leads to a very homogeneous and flat Universe. Let us imagine that prior to inflation we are in a region where space is very curved, that is, with a small radius of curvature. Inflation will produce an exponential growth of the region and its curvature radius at the end of which space will appear flat. The inflationary process also explains the observed homogeneity. Even if the Universe had had very different temperatures in non-causally connected regions, what we are observing today is the expansion of a minuscule portion that had already achieved thermal equilibrium. Therefore, according to inflation, our current visible Universe only corresponds to a very limited region of the primordial Universe.

One may be left with the impression that the proposed solution achieves its goal excessively and at the end of the process the Universe would be so perfectly homogeneous that the small differences in temperature and density we observe could not be explained. Without these inhomogeneities the hot and cold observed regions would not exist. Since they are the seeds of the regions dense in matter content where galaxies and stars are formed, the latter would not have arisen and without them we are again facing a Universe that cannot harbor life. The presence of inhomogeneities is predicted with remarkable precision when one incorporates into the inflationary corrections stemming from quantum mechanics. Let us recall that in the latter theory one has the uncertainty principle that predicts that fluctuations in the energy can take place. The bigger the fluctuations the shorter they last and in particular they allow for the creation and successive annihilation of pairs of particles. The attentive reader would question that quantum mechanics is about the small and the kind of fluctuations it involves cannot be the ones that give rise to fluctuation in the density and temperatures and are related to the formation of galaxies. Once again inflation resolves this point. Let us imagine the Universe during inflation as a balloon, initially small, that expands enormously. If someone had penned a small, almost imperceptible, drawing on the balloon, the drawing would become evident as the balloon grows. Similarly, the quantum fluctuations of microscopic nature at the beginning of inflation are amplified. This leaves an imprint on the cosmic microwave background, and its distribution coincides spectacularly well with the observed one. After the inflationary period, due to the amplified fluctuations the density of matter will vary slightly from place to place. Due to the effect of the gravitational forces the denser regions will contract, forming the galaxies.

The original model of inflation proposed by Guth requires certain initial conditions in order to take place. To avoid the introduction of special initial conditions Andrei Linde (1982) and Alexander Vilenkin (1983) found a form of inflation that does not require special initial conditions. The idea of this type of inflation is that it is due to a field that undergoes quantum fluctuations. The larger the field the stronger the inflationary process. Since the field is subject to quantum fluctuations, its intensity will change in different times and places. In the regions where the fluctuations make the field large the inflationary process is stronger and the exponential growth of space will be more intense than in other regions. In other words, the few regions where the field becomes very large will create bigger expansion and will therefore become the dominant ones. Most of the Universe will continue in inflationary expansion, that is, will keep on doubling its size in fractions of a second whereas in other regions, like the one we are in, inflation would have stopped. This process is basically eternal because new fluctuations would induce new exponential expansions. Summarizing, the scenario is that of a Universe that continues inflating indefinitely with regions where inflation has ceased like that of our observable Universe. Although the inflationary scheme of Linde and Vilenkin is not completely confirmed it leads to a vision of the Universe with a size that significantly exceeds the observable Universe and that in fact would contain a countless number of Universes separated by inflating regions. In fact it is not even necessary to think that the Universe underwent an epoch prior to inflation, it could be that regions undergoing inflation and creating in certain places Universes like ours have always existed. Linde calls this model eternal chaotic inflation. This description offers the possibility that instead of having one Universe we have infinitely many ones separated by inflating regions. This is known as a multiverse.

13.4 The Anthropic Principle

At least on Earth conditions for the development of life have occurred and through the evolutionary process it has led to beings capable of constructing representations of the Universe like the one cosmology provides. The representational ability of humans is perhaps our most distinctive characteristic. It manifests itself in many ways even in the most primitive stages of human development. To mention some examples (Jonas 2001) its ability to represent images like those in the Altamira caves in Spain, which are initial manifestations of our ability to represent, that in more advanced phases led to mathematical language. It is surprising that the evolution of the Universe ended generating beings capable of constructing scientific representations of that very development. More general, in a Darwinist scheme where humans are just another life form with a particular degree of development of certain abilities, it is surprising that the Universe was capable of generating life. Brandon Carter (1974) was probably the first in making explicit the reasons to be surprised. He asked the question "Suppose the laws had been different from what they actually are... what would the consequences be?" Specifically, Carter was wondering about what would happen to the existence of life if the physical laws, in particular the values of the fundamental physical constants like the masses or charges of elementary particles, took slightly different values than in our current Universe. Which small changes would suffice to make life impossible? In other words, he discovered that the fundamental physical constants seem fine tuned to generate a universe hospitable to life. To characterize this fine tuning the physical laws with the existence of of life Carter coined the term *Anthropic Principle*. This was followed by many papers and a strong debate about the meaning and implications of such principle. For many it lacked any explanational power.

Carter defined two types of anthropic principle. The *Weak Anthropic Principle* (WAP) refers to the fact that our spatio-temporal placing in the Universe must necessarily be privileged in order to exist as living beings capable of thought. This is quite evident because a) the existence of life requires favorable conditions —certain temperatures, the existence of certain substances like water— that do not exist in an arbitrary point in the Universe and b) the Universe evolves and was not suitable for live when it was very young or when it becomes much older. The belief stemming from the Copernican revolution that we live in a corner of the Universe very similar to any other was not entirely correct. Although in our galaxy there are hundreds of billions of stars, many of which are similar to the Sun with planetary systems, and there are a similar number of galaxies, planets like the Earth with all the conditions for a phenomenon as complex and delicate as life to develop could be rare. It could even be that the Earth is the only such planet. The WAP emphasizes the need of privileged conditions like those on Earth to allow the existence of observers. However, estimates of habitable zones around other stars, the great abundance of elements needed for life like carbon and hydrogen and of organic molecules in interstellar clouds, along with the discovery of hundreds of extra-solar planets and new studies of extreme habitats on Earth suggest there may be many more habitable places in the Universe than considered until very recently. Studies of exoplanets estimate that up to 25% of the stars may have planets in habitable regions, that is, regions like the one the Earth occupies, in which the radiation received suits the needs of life. Recent estimates set the number of Earth-like planets in our galaxy that orbit around stars similar to our Sun in 10 billions (Cassan 2012). Of course, there are additional conditions needed to allow the existence of life in other planets, so we do not have a reliable estimate of their number.

The *Strong Anthropic Principle* (SAP) states that the Universe, and therefore the fundamental constants it depends on, must be the required ones for life to be possible at some stage of its history. It is clear that these statements, without an explanation of the reasons that lead to the fundamental constants being tuned in such a special way are not very satisfactory. Beyond the controversy, the developments of the last few decades have only added evidence that the existence of life in a Universe depends among other things of a very fine tuning of the physical constants in order to ensure a series of phenomena that as we shall see later are key to ensure the existence of life based on the physics and chemistry of carbon. At the same time it became to be understood that many laws and constants that appeared fundamental are valid only at low energies and could lead to different behaviors in other regions of the Universe. We will devote the last part of this chapter to explain some of the "coincidences" that must occur in the values of the fundamental parameters in order for life to be possible and we analyze possible explanations for such a degree of fine tuning.

13.5   A bio-friendly Universe

As we have seen, there is a growing conviction that the observed values of a large number of dimensionless physical constants, like the fine structure constant that depends on the charge of the electron and other constants, seem to take values finely tuned to permit the formation of matter and the appearance of life. A small increase in the strong interaction, for example, would have converted all hydrogen existent into helium. Without hydrogen, stars with very long lives like the Sun would not exist. Water would not exist either and therefore two crucial requirements for life would not be met.

Paul Davies lists three basic requirements that have to be satisfied: a) The laws of physics should permit stable complex structures to form; b)The universe should posses the sort of substances, such as carbon, that biology uses; c) An appropriate setting must exist in which the vital components come together in the appropriate way. In this section we will follow closely his presentation of this topic (Davies 2006).

In what concerns the first condition, most of the hydrogen and helium that were formed in the first stages of the life of the Universe long before stars were formed. The relative abundance is determined as we mentioned by the strength of the strong interactions. It is crucial for life to have more hydrogen than helium so long lived stars form in order to allow for the times required for the development of life and its evolution to take place in stable conditions of temperature and radiation. The other elements that were created in the stars, which behave like gigantic nuclear reactors. Stars obtain most of their energy from converting hydrogen to helium. Let us recall that hydrogen has a single proton whereas helium has two protons and two neutrons. Free neutrons have a mean life of a few minutes. As they are not stable, by the time stars were formed they had disappeared and only protons were available. But protons have electric charge and repel each other. In order to form a helium nucleus one needs to get the protons to within a trillionth of a centimeter of each other. In the star interiors the pressures are so large that the protons become close enough to each other to allow for the very short-ranged nuclear forces to come into play and counterbalance the electrical repulsion. In order to form helium one not only has to get four protons at a short distance from each other, one also needs to transform two protons into neutrons. Up to now the phenomena only involved three of the interactions. Gravity is responsible for the formation of stars with very dense interiors. Electromagnetism accounts for the repulsion of protons. The very short-ranged nuclear force attracts the protons. To create the neutrons one needs a fourth interaction, called weak interaction. Its weakness slows down the creation of neutrons, which is crucial in order for the helium nuclei to form at a slow pace and the stars burn their fuel slowly, which allows them to be long lived, a key requisite for the development of life. Therefore the star processes, their longevity and the formation of helium, which is the first step towards the formation of the other chemical elements depend on a fine tuning between the intensities of the four fundamental forces.

In sufficiently massive stars the process of nuclei formation for the heavy elements continues. But creating nuclei heavier than helium's also requires a fine tuning among the fundamental forces. Indeed, at first sight all paths for creating heavier elements seem blocked, both for the formation of lithium and beryllium since they involved forbidden or very unlikely processes. This led to the study of processes where the element created was carbon. For that one would need three helium nuclei to fuse to give a carbon nuclei with six protons and six neutrons. Again the process appeared very unlikely, which leads us to the second requirement for life that Davis discussed.

The second requirement was that the Universe should possess the sort of substances, such as carbon, that biology uses. The problem appeared unsolvable until Fred Hoyle in the early 1950's found an unexpected solution based in the idea of resonance. When two objects like the string and the body of a violin resonate, the oscillations in one of them stimulate oscillations in the other with equal frequency, amplifying the amplitude of such oscillations. This phenomenon has multiple applications, it is for instance used when tuning an old regenerative receiver of the 1920's for a particular radio station. In order to do so we make the internal circuitry in the radio receiver at a natural frequency that enters into resonance with the waves incoming from the radio station, amplifying it. In order to use the idea of resonance to explain the formation of carbon in the stars Hoyle argued that the notion of resonance applies to Schrödinger's equation, that is quantum waves, and these could supply the missing element in the construction. Let us suppose that two helium nuclei fuse together giving rise to a beryllium nucleus. The resulting beryllium is very unstable and disappears very quickly, which prevents it from merging with the third nuclei and produce the carbon. Hoyle's hypothesis was that the system composed of helium/beryllium could enter into resonance with an excited state of the carbon nucleus extending the lifetime of the beryllium enough so the fusion of the helium and beryllium could be completed. When Hoyle proposed his idea there was not much known about the excited states of carbon. However, spurred by the work of Hoyle nuclear physicists could verify that indeed such states existed and that the resonance enhanced the lifetime of the beryllium ensuring that the third helium could combine with it. Only due to the existence of such excited state at the exact required energy for the carbon resonance the heavier elements, and therefore life, can exist. The value of the energy of that excited state depends very finely on the intensity of the nuclear interactions involved: the strong and the weak forces. A variation of 1% in the value of the nuclear force would be enough for the resonance and therefore the creation of carbon not to occur (Oberhummer *et al.* 2000). The number of restrictions on the fundamental parameters of physics is much larger, but let us take a look at an example of the third type of requirement.

The third requirement was that an appropriate setting must exist in which the vital components come together in the appropriate way. In particular a mechanism must exist to ensure that the elements produced in the stars, in particular carbon, disseminate and reach us. A frequent mechanisms to enable this in very massive stars is that they explode forming supernovae and ejecting into space a good portion of their material. The mechanism of formation of supernovae called "type II" depends crucially on the weak nuclear force.

When the fuel constituted by the light elements runs out and the mechanism of nuclear fusion stops producing the heat of the nuclear energy that supplies the pressure needed to sustain the star, its nucleus starts to contract very rapidly in a cataclysmic implosion. This in turn rises the temperature and a phenomenon of conversion of protons to neutrons takes place and the nucleus of the star turns into a hyperdense mass of neutrons called neutronium. Each time a proton turns into a neutron it absorbs an electron and produces a neutrino as a result of the weak nuclear interactions. An enormous sea of neutrinos develops in the interior of the star. In normal conditions neutrinos, which have a negligible mass and interact only through the weak nuclear interactions, traverse matter without interacting. However, in the very massive stars nuclei neutrinos remain blocked without being able to exit due to the very high density of the stellar nucleus. This implies that as more neutrons and therefore neutrinos are formed, their pressure increases and eventually an outward shock wave is produced that drags out all the material of the exterior layers of the star. The ejected neutrinos are actually detected by underground instruments on Earth. It has been experimentally verified that each time a supernova close to Earth forms, the visual observation of the supernova is accompanied by the detection of a strong flux of neutrinos. If the weak nuclear interaction were weaker the collapse of the stellar nucleus could not stop and there would be no explosion. If it were stronger the explosion would happen earlier and would eject less material.

The evidence for the fine tuning of the fundamental constants does not end here. In fact, it can be said that the fine tuning problem has turned into a problem in contemporary physics without the need to introduce the discussion of the requirements for the existence of life. Again, only to mention an example we will discuss one of the central problems of contemporary physics: the problem of the cosmological constant.

We saw that dark energy produces a repulsive force that makes the galaxies recede at an accelerated pace. The dark energy can be due to the existence of a scalar field, as it happened during the inflationary process. But, as Einstein himself had noted, it appears as an extra term in his field equations for gravity known as a cosmological constant, that in principle can take an arbitrary value. In the last decades, studying the vacuum energy of quantum fields a highly problematic prediction was obtained: the the dark energy density had to take a value that is some 10^{120} times bigger than observed. That is one followed by 120 zeros. This is perhaps the largest discrepancy between theory and experiment ever[1]. The problem, which was encountered decades ago, has not found

---

[1]Some have argued that these calculations are naive and a more rigorous calcula-

a satisfactory solution. One could assume that since Einstein's theory admits arbitrary values of the cosmological constant, one could use it to cancel out the contribution of the dark energy arising from the vacuum of the fields. What is difficult to understand is how that arbitrary value cancels almost perfectly the large vacuum energy, reducing it to a value 10^{120} smaller than the natural one. More serious yet is the fact that only because the observed dark energy is so small galaxies could form and create an environment friendly to life. If the dark energy were only ten times bigger the Universe would expand so quickly that it would prevent the formation of galaxies. Only that perfect cancellation between the original cosmological constant of the Universe and the one resulting from the energy of the vacuum of the fields ensures the existence of a universe populated by galaxies —and therefore suitable for life— as the one we observe. The problem of the fine tuning of the physical constants has become a source of shame for contemporary physics and needs to be explained.

We will devote the next chapter to analyze possible solutions to this problem.

---

tion actually shows that the vacuum energy of the fields is way too small to account for the dark energy. In any case the problem persists. See Hollands (2013).

Chapter 14

The Multiverse and beyond

The modern concept of a multiple universe or *multiverse* has become very popular in the last decade. This is due to several factors, including a few we have already mentioned. One of the most important ones results from the success of the inflationary hypothesis in explaining the cosmic microwave background radiation which we have already mentioned. In particular we saw that inflation predicts precisely the fluctuations in temperature observed in the cosmic background radiation and the fluctuations in density that explain the formation of galaxies. Cosmologists studying inflation stumbled upon the idea of multiverse while noting that the inflationary process, once started, would very likely continue indefinitely. Inflation may stop in certain regions after some time but it will continue expanding the Universe in others.

Today we know that the visible Universe resulted from the expansion of a region that exploded 13.8 billion years ago. As the galaxies recede with speeds that grow with distance, the most distant objects recede at speeds close to that of light. The observable Universe for an observer on Earth contains only part of the matter that expanded after the Big Bang. The observable Universe for an observer in a distant galaxy will contain galaxies not observable from Earth. In other words, it is a sure thing that the Universe extends beyond what we see. What the inflationary hypothesis of Linde and Vilenkin adds is that the portion of the Universe we cannot see is essentially infinite and while in our observable portion the Universe expands slowly, in other regions it continues to expand exponentially.

In this scenario, in spite of inflation stopping in this particular region, it continues its crazy expansion in many others: "one inflationary bubble sprouts other inflationary bubbles which in turns produce other inflationary bubbles." (Linde 1986). Since the process continues indefinitely, it ends up producing an infinite number of universes.

This opens the possibility of explaining why we can have fundamental constants fine tuned to explain the existence of life or why we have such an unnatural value of the cosmological constant. If in different regions the fundamental constants were different then there could exist universes suitable and unsuitable for life. It would not be strange that we are in one of the suitable ones since we could not exist in an unsuitable one. The problem that remains to be explained is why the fundamental constants change from one region to another. At least for some of the models of inflation this is actually possible.

The recent discovery of the Higgs boson in the Large Hadron Collider (LHC) at CERN has confirmed that the masses of the elementary particles like the electrons or the quarks do not take arbitrary values without explanation. On the contrary, the Higgs mechanism explains how these particles acquire mass. Briefly speaking, the idea is the following: electrons and quarks start massless, they acquire their masses through interactions with the field of the Higgs boson. The mass they acquire depends on the intensity of the interaction of each particle with the Higgs boson. Quarks interact more strongly than electrons and therefore have larger masses. The masses do not only depend on the strength of the interaction with the Higgs boson but also on the intensity of the Higgs boson field. Such intensity could be different in different regions of the Universe, and that would give rise to changes in the masses of quarks and electrons. Although these changes would happen *en bloc* in the sense that the relation between the masses of the electrons and quarks would remain constant, the masses of some particles like protons and neutrons not only depend on their components but also on the form in that those components interact among themselves to form the proton. In other words, the changes in the masses of the proton and neutron will not have constant relations with the masses of the electrons and quarks. A change in the relation between the mass of the proton and the electron would change how stars work, perhaps making impossible the creation of heavy elements and therefore life.

But changes in the values of the masses are not enough to have a life friendly Universe. We saw that it was crucial that the constants associated to the fundamental forces: gravitational, electromagnetic, weak and strong, all maintain certain very precise relationships in a bio friendly universe. The multiverse that results from a Linde–Vilenkin type of inflation would not suffice to explain how the fundamental forces emerged so synchronized.

At the beginning of the 21st century the notion of multiverse had a new boost, this time coming from string theory and the attempts to solve the problem of the absurdly unnatural value of the cosmological constant. Let us recall that string theory, considered one of the most promising candidates to be "theory of everything" replaces the notion of point particle by a string moving in an eleven dimensional space. The vibrational modes of the strings are associated with the different particles. The initial hope of string theorists was that it would explain why the fundamental parameters like Newton's constant, the cosmological constant and the coupling constants of the other forces take the values they do. This goal, that many physicists consider the Holy Grail of a physical theory, is far from being achieved. In fact, pursuing it revealed that the theory has a gigantic number of stable configurations of space-time. Estimates of that number lead to a "landscape" that contains of the order of 10^{500} universes with different physical properties. In particular the cosmological constant could take values in many cases as improbable as the one we observe.[1]

Linde and Leonard Susskind are two of the main enthusiasts of this conception of "megaverse" or "landscape". It would allow to think that the various bubbles of the inflationary cosmology could correspond to one of the 10^{500} vacua allowed by the string landscape. Given that the number of different vacua is finite and the number of possible bubbles infinite, each vacuum could repeat an infinite number of times, for instance reproducing a Universe like ours. The hypothesis about the existence of the multiverse is not easy to accept for many physicists. As Nobel laureate Steven Weinberg says, if the multiverse exists, "the hope

---

[1] An alternative proposal to select the value of the fundamental constants was put forward by Lee Smolin (1999) who argues that a new universe is formed inside every new black hole that emerges. In the resulting new universe the fundamental constants may take slightly different values and a process of "natural selection" of the constants towards values that maximize the production of black holes takes place.

of finding a rational explanation for the precise values of quark masses and other constants of the standard model that we observe in our Big Bang is doomed, for their values would be an accident of the particular part of the multiverse in which we live." The physicists' ambition for understanding the universe would be severely limited if we can only say with certainty that we are in one of the 10^{500} possible universes and that we landed in it by chance.

That something is bad for theoretical physics does not mean it is false, but since we have turned somewhat speculative in these last sections, we should distinguish well established science and what is merely speculation, even if it is speculation by distinguished physicists. To begin with, the notion of multiverse we presented depends on inflation occurring in a specific way. There are indeed many models of inflation and only some of them give rise to eternal inflation like Linde's does. Although the current cosmology allows us to suggest, based on the information from the cosmic background radiation, that the Universe underwent an inflationary period, it is difficult to determine if the specific inflationary process that occurred led to the creation of multiple universes, that is, the multiverse. Moreover, the notion of landscape, that is, that different universes correspond to different fundamental constants depends on the validity of string theory. At the moment there is no experimental confirmation of string theory. It can only be considered an attractive proposal to unify the fundamental theories and interactions. Lastly, the multiverse hypothesis appears impossible to verify experimentally. We could only have indirect evidence that could stem, for instance, of an eventual future experimental verification of string theory[2].

It has been proposed that it would be possible to explain the characteristics especially receptive to life of our Universe studying the subset of habitable universes among the enormous variety allowed by the landscape. It would be good if one were able to show that within those universes capable of harboring life and observers capable of knowing it ours is in some sense typical and does not require additional fine tuning

---

[2]One should mention here that superstring theory is a particular case of supersymmetric theories and up to now there is not any experimental confirmation for the existence of supersymmetry. Direct confirmation would require detection of superpartners of the particles of the Standard Model, for instance in Large Hadron Collider (LHC). In fact supersymmetry predicts that for each fermion would exist a boson and vice-versa. Up to now no evidence for supersymmetry was observed. This sets limits on the energies of the superpartners.

of the fundamental constants. However, it appears reasonable to assume that among the universes capable of harboring thinking life the more frequent ones will be those in which this situation happens exceptionally. Indeed, one would expect that the parameters, that in order to harbor life should take values within certain ranges, are not in addition selected much more finely for our Universe to have the optimal conditions for life. If that were the case the anthropic principle would not be enough to justify our existence in a Universe friendly to life and an additional fine tuning will be needed. The fact that we are in a Universe that contains beings capable of knowing it would not justify that life is an extended phenomenon and typical in our Universe with millions of possible planetary systems capable of generating life.

Coming back to the question of what can explain that our Universe is life-friendly: does the multiverse resolve the problem? Be it that the Universe is a single one or multiple ones, the fact that it is bio-friendly does not depend only on the values that certain parameters take. It depends strongly first and foremost of the great elegance and creative potential of the physical laws that govern it and that are the same in the whole Universe. It depends on the fact that quantum mechanics is as it is, of the geometric and dynamical nature of space-time predicted by the general theory of relativity (or the eventual theory that unifies it with quantum mechanics, be it string theory or other). Those laws, not only universal but "multiversal" are who define the order capable of ruling a world that is life-friendly. Those are the laws that marveled Einstein for example, when he observed that "Everything is determined, the beginning as well as the end, by forces over which we have no control. It is determined for the insect, as well as for the star. Human beings, vegetables, or cosmic dust, we all dance to a mysterious tune, intoned in the distance by an invisible piper."

Although it is not made explicit on many occasions it also depends on the ultimate nature of the Universe that the laws described, of the ontology that results applicable. The nature of the Universe depends on the description and the notions like that of field, state, event or physical system that constitute the ontology of our Universe. To understand life and consciousness without understanding what the fundamental laws refer to is an impossible task. Ultimately it is those laws and the nature of the fundamental entities ruled by them who are bio-friendly.

It does not appear to make sense to ask why these laws and not others. Is it maybe possible to demonstrate that they are the only possible laws? Based on what axioms? It appears inevitable to enter a vicious circle in that each explanation requires an additional one. We are facing the known history of the Universe and the turtle that Davies returns to in his book. While a famous speaker[3] gives a talk about the nature of the Universe someone from the audience interrupts to tell him that he knows how it works: the Earth lies on a giant elephant that stands on a giant turtle. When the speaker asks what does the turtle stand on, the member of the audience responds: that is no problem, it is turtles all the way down! Putting it in more serious terms we should recall the so-called Leibnitz' Modal Argument: "the existence of a contingent reality can only be ultimately explained through a cause whose existence is in itself necessary. However, something whose existence is in itself necessary is something whose existence cannot depend upon anything else but itself, its own nature." Einstein wished precisely to understand whether "God had any choice in the creation of the world." If the answer were negative then the laws of the Universe would be necessary and therefore the only ones that would make possible its existence. The most economical hypothesis when facing these ultimate questions would be to have a Universe with necessary laws and contingent behaviors. But it is clear that the decision made at this point is undemonstrable: either a single Universe with laws and parameters given, the multiverse where the laws are unique but all possible selections of parameters are instantiated or a god created Universe.

Implicit in this quest is the principle first laid out by Spinoza (1663/2007) and elevated to a principle called the "Principle of Sufficient Reason", that states: "Nothing exists of which it cannot be asked, what is the cause (or reason) [causa (sive ratio)], why it exists." To have elevated to a logical principle this inexhaustible need of explanation may lead —given the lack of answers about the inexplicable hospitality of the Universe and its physical laws— to the feeling that we live in an absurd Universe. Nevertheless, in our understanding the Universe does not appear as absurd when this need to which mankind feels drawn to in a natural way —for example to explain why these laws and not others— is questioned. The notion of absurd appears when the search for assigning it meaning fails.

That thinker of the absurd, Albert Camus (1942/1955), describes this search of meaning in the following way: "If man realized that the

---

[3]The story has been told many times, dating back to the mid 1850's.

universe like him can love and suffer, he would be reconciled. If thought discovered in the shimmering mirrors of phenomena eternal relations capable of summing them up and summing themselves up in a single principle, then would be seen an intellectual joy of which the myth of the blessed would be but a ridiculous imitation. That nostalgia for unity, that appetite for the absolute illustrates the essential impulse of the human drama But the fact of that nostalgia's existence does not imply that it is to be immediately satisfied. For if, bridging the gulf that separates desire from conquest, we assert with Parmenides the reality of the One (whatever it may be), we fall into the ridiculous contradiction of a mind that asserts total unity and proves by its very assertion its own difference and the diversity it claimed to resolve. This other vicious circle is enough to stifle our hopes."

Camus lays out two basic requirements that we have tried to establish in this book: the first one is the quest for an order behind phenomena: "eternal relations capable of summing them up and summing themselves up in a single principle" and second what Camus states first, the aspiration that the Universe not be a mechanism estranged from human concerns; of a Universe that could also love and suffer. Notice that Camus does not aspire to an exhaustive explanation that does not leave anything unexplained but just to "a unique principle" for the natural and the human.

Worried about saving "the gulf that separates desire from conquest" due to an excess of sensitivity and lack of patience and in good measure under the influence of the nihilism of the time, Camus renounces to insist in the quest of answers in science. He recognizes this explicitly when he says that "Yet all the knowledge on earth will give me nothing to assure me that this world is mine. You describe it to me and you teach me to classify it. You enumerate its laws and in my thirst for knowledge I admit that they are true. You take apart its mechanism and my hope increases. At the final stage you teach me that this wondrous and multicolored universe can be reduced to the atom and that the atom itself can be reduced to the electron. All this is good and I wait for you to continue. But you tell me of an invisible planetary system in which electrons gravitate around a nucleus. You explain this world to me with an image. I realize then that you have been reduced to poetry: I shall never know. Have I the time to become indignant? You have already changed theories. So that science that was to teach me everything ends up in a hypothesis, that lucidity founders in metaphor, that uncertainty

is resolved in a work of art. What need had I of so many efforts? The
soft lines of these hills and the hand of evening on this troubled heart
teach me much more. I have returned to my beginning. I realize that if
through science I can seize phenomena and enumerate them, I cannot,
for all that, apprehend the world. Were I to trace its entire relief with
my finger, I should not know any more. And you give me the choice
between a description that is sure but that teaches me nothing and
hypotheses that claim to teach men but that are not sure."

Having abandoned the quest of the ultimate turtle that holds the
world we will concentrate in what follows in the question about the
meaning in a naturalist context as formulated by Albert Camus.

Part IV

Religious Naturalism

Before moving into the final chapters of the book, we would like to stress that they clearly have a different character than the ones preceding them. Here we enter into significantly speculative terrain. Our intention is not to present a definitive philosophical system. Such a study would require many new developments not only at the philosophical level, but also at the scientific one, including developments in biology, chemistry and physics. Our goal here is to stimulate the reflection about some issues raised by the recent scientific results and open a dialogue about their ethical and spiritual implications. We consider that the issues raised are of great interest and should spark further debate among philosophers and/or physicists.

Chapter 15

The scientific roots of nihilism and its overcoming

15.1 Introduction

Our thesis is that since the beginning of the 20th century, a series of scientific changes take place, characterized by the a set of fundamental conceptual revolutions. We have dedicated a good portion of the previous chapters to them and showed how we have entered a new era of understanding of the world. It surprises us as something completely new that shatters the false contradictions among which philosophical thought has struggled for centuries. We refer to the developments of the fundamental physical theories: general relativity and quantum mechanics, but also to the remarkable progresses in the understanding of life. In them, the cell taps into the genetic information according to its needs and those of the environment in the epigenetic processes. Also the advances in the understanding of complex systems with top-down causation and the incredible progresses in astrophysics and cosmology that take us close to the first instants after the Big Bang and reveal a world much more hospitable to life that one would have expected from the mechanist paradigm.

These conceptual revolutions break a tendency of 300 years in which

the striking results of a science that was still primitive and the decay of
the traditional religious visions led to a loss of certitude and a nihilist
vision of mankind in its relationship to God and nature.

15.2   Nihilism in 19th century science

There is no doubt that the role played by formal religious systems in
the Western World has suffered a significant decay in the last centuries.
The root causes and origins of the phenomenon are multiple and in some
cases precede the birth of modernity or are associated to specificities of
the very Judeo-Christian religious tradition.

Towards 1650 the radical illuminists, deist or atheist, reject the
Judeo-Christian tradition, its idea of the creation, the faith in the intervention of a providential God in human affairs, the miracles and the
rewards or punishments after death. At a social level, they reject the
authority of the Church and the existence of an aristocratic order of
society consecrated by the Divine Will. Instead they profess a great
reverence for the experimental sciences and mathematics and they pro-
claim that mankind in the future will have as only objective their own
happiness in life. It is not a movement limited to intellectuals or the
most educated ones. The crisis in the ideas and trends impacted quickly
on society as a whole. Towards the end of the 17th century even illit-
erate people had been infected by the new ideas and the masses were
beginning to doubt the existence of hell and sorcery and to recognize
the use of superstition as a form of social and intellectual domination.

Modern thinking, based on the physical sciences, was developed
starting from a mechanist notion of matter deprived of all the properties usually attributed to life. The Universe is conceived as an infinite
space where inanimate masses move under the action of forces that
obey immutable laws according to the laws of inertia. To reach this
conception the first physicists —Galileo is a paradigmatic example—
had to make a remarkable effort of abstraction that had to eliminate
the vital aspects together with feelings and sensations from the physical
description. The latter was given in terms of a mere extension of the
mathematical formalism. Thus, it constitutes in a central and paradigmatic aspect of reality since it was, in the end, all that was cognizable.
Concerning life, the remaining task was to explain it in purely mechanical terms. It should be conceived as an improbable accident that maybe

only happens in this corner of the Universe and must be explained reducing it to a mere mechanism composed of masses and forces. Those are the demands of the chosen ontology and therefore the pending task for the biological researchers once that conception of the physical nature of things has been adopted.

But it is the very possibility of knowledge, of a mind that thinks and conceives the physical explanation, what remains outside the ontology postulated from the discoveries of the recently created new physics of the time. It is the very discovery which allows mankind to understand and rigorously predict the behavior of matter without spirit what leaves us to recognize that such a thing is only possible due to the ability of human thinking. This leads to the dead end alley of dualism. Once both substances —one related to matter the other with mind— are conceived as independent and defined in terms of incompatible attributes, the question of interaction of the substances arises, on which the possibility of any knowledge depends. Descartes' dualist position bifurcates the world in an apparently definitive way between a material sphere that obeys rational and intelligible laws formulated in mathematical terms and the thought. The latter is limited to the immediacy of its self-consciousness as an inexplicable and unintelligible residue totally disconnected from a material world paradoxically accessible to rational thought. Since then, the world bifurcates into two contrasted positions: materialism and idealism, which mutually attempt to reduce each other to non-existence. If nature is no more than, as Descartes would put it, *res extensa*, the human being, the only one endowed with thinking, remains separated by an abyss from the rest of existence due to their consciousness. As Hans Jonas observes "his consciousness only makes him a foreign in the world, and in any act of true reflection tells of this stark foreignness."

The other great development of science in its first stages that led to nihilist positions was given by the 19th century evolutionism. The development of life is stripped of any plan or purpose other than self-preservation. This reminds the role of inertia as a blind impulse that rules and keeps in motion particles. The development of life, led by the enormous variability of its conditions of existence is totally contingent and lacks any plan or previous design. The randomness in the life conditions in the most diverse environments and individual adventures of every living being is antipodal to the mechanical laws that supposedly rule its behavior. The evolutionist vision transforms all teleological behavior in mere appearances. Thus, for example, orchids adopt infinite colors and forms without any other purpose than that of optimizing their reproductive capacity in the environments they develop (Darwin 1862/1977).

Nietzsche (1888/1968) discusses under the title "Collapse of Cosmological Values" how what he calls *extreme nihilism* arose in Europe. He characterizes the nihilism as a psychological condition that manifests itself in three ways:

1) *A world lacking any purpose* This form of nihilism appears when the conviction that the development of the world has a purpose ends in disillusion. When there exists the conviction that the world has an end or destiny to which it leads humans can feel protagonists of that process. Or at least taking part in it they can find a goal for their lives and feel that they have meaning. The purely mechanist conception and the elimination of all teleological conception of nature strips it from any reference to goals and lacking its development of purpose it does not provide justification for any human purpose. Such a world stripped of senses manifest itself mainly for its quantitative enormity, its almost infinite extension or the violence with which it occasionally expresses its power in cataclysmic events like earthquakes of tsunamis on the Earth or a supernova explosion at a cosmic level. Jonas says, referring to this situation created by the first developments of science: "if the world has anything at all to tell of the divine, it does so through this property: [its magnitude—that is the purely quantitative properties like extension or energy-] and what magnitude can tell of its power. But a world reduced to a mere manifestation of power also admits towards itself— once the transcendent reference has fallen away and man is left with it and himself alone-nothing but a relation of power, that is, of mastery." In the primitive Darwinist vision human nature is reduced to another organism that deploys its aptitude to survive and the development of life ends up being nothing but the manifestation of reduces everything to a *will to power*. With this, human actions are completely stripped of ethical fundamentals. A human being that is one more animal, to whom he wishes to attribute free will which is denied to the rest of the living beings. A foreigner, only thinking being in a blind and mechanical world.

2) *The fracture between a mechanical and indifferent world and a human being that feels and suffers.* Nietzsche recognizes as a second form of nihilism the loss of the belief that binds the human beings to a great whole that transcends our personal experience and enables us to

devote ourselves to a common welfare. When the notion that we belong to a whole greater than the parts that gives meaning to our existence is lost, been this whole: our nation, the entire cosmos or God, "the highest values become devaluated." When "God is dead" and nature behaves as a pure automaton: enormous, mechanical and inhospitable to life. The latter is no more than an improbable accident in an absurd Universe totally indifferent to human worries. Humankind inhabits it endowed of a sensibility and conscience completely incompatible with its mechanical nature. The supreme absurd of human beings in their ability to react with anguish and desperation, of feeling a profound rejection of this situation, when according to this view they would be just another mechanism that is blind and indifferent. This is the fundamental anguish felt by Camus and that he described in The Myth of Sisyphus (1942/1955).

3) *The disbelief in any transcendent reality.* The third form of nihilism described by Nietzsche stems from recognizing that any attempt to erect an ideal world above the physical one is unsustainable and only responds to our desires without any real justification. Nishitani (1990) describes it in these terms: "the final form of nihilism arises, which includes the disbelief in a metaphysical world... In this standpoint one recognizes the reality of becoming as the only reality, and forbids oneself any kind of scape to other worlds and false divinities." It is a disbelief in the very possibility of the existence of a world of things in itself, of a real world. This position has strong resonances with *Logical Positivism*, which rejects all possibility of accessing things as they are and that affirm that science is the only form of knowledge. More precisely, only the statements verifiable either by deductive logic or direct observation would be cognitively meaningful. Ethics and metaphysics are considered as part of the unscientific discourse and unfit for the philosophical practice, whose task is nothing but to organize knowledge, not develop new knowledge. This position, although apparently attractive, resulted inconsistent and unsustainable. As John Passmore (1967) expressed it, positivism is "dead, or as dead as a philosophical movement ever becomes". And even a positivist like A. J. Ayer recognized in a television interview that the most important defect of positivism was that "nearly all of it was false."

15.3 Revision of nihilism from the perspective of contemporary science

Nietzsche's nihilist vision must be revised in light of the scientific knowledge of the last century that, as we have shown in previous chapters, leads to a considerably less negative vision. Let us review these changes in perception in each of the mentioned aspects. In this chapter we will limit ourselves to the analysis of the first form of nihilism: the one that does not recognize any purpose in the world.

The notion of purpose applied to the physical world arises from an implicit extrapolation of the concept of intentional action in the case of human being where the means are subject to obtaining an end. Science studies natural phenomena in terms of deterministic or probabilistic laws that establish conditions for the development of processes on the basis of the knowledge of the initial conditions. One speaks of teleology when one considers that the natural order obeys an organizational principle that transcends the mere regulations of the processes by the laws. Aristotle is the greatest champion of finalism of antiquity. Inspired by the observation of the organic world, he assumes that the finalism also applies to the inorganic world where things move spontaneously if nothing interferes with them to their "natural place". Aristotle assumes that purpose without any sort of conscious elaboration can exist. That is, that purpose forms part of the natural order without the need to resort to any supreme being. The Aristotelic teleological vision, that attributes a central role to finalism in the scientific explanation has been eradicated from modern science whose main aim is to establish rules —laws— that relate initial conditions with the later development of phenomena. They establish limitations that determine partially or completely the future behaviors, specifying possible future behaviors and their probabilities.

Although science eliminated the final causes that justify the evolution of a system to satisfying a purpose, it is practically impossible not to refer to a countless number of objects or institutions that have certain purposes. In many cases the reference arises in connection with a product of the design of the object. For instance, matches are created to start a fire. However, its purpose is not that of the object but that of the manufacturer. The use of the notion of purpose does not imply necessarily the existence of someone or something that conceives the end before seeking the means to realize it. An example along these lines is the concept of teleonomy associated to the apparent purposefulness of organs and functions developed in living organisms through the action of natural laws and evolutionary processes. The very word makes reference to this double aspect combining the Greek words "telos" (goal or purpose) with "nomos" (law). One of the proponents of its use was Jacques Monod (1970/71) who observed that "Rather than reject this [goal-directedness] idea (as certain biologists have tried to do) it is indispensable to recognise that it is essential to the very definition of living beings. We shall maintain that the latter are distinct from all other structures or systems present in the universe through this characteristic property, which we shall call teleonomy." Every time that a species adapts due to changes in the environment or to move to another environment that up to that point appeared as hostile there are changes that adjust to an end and "goal-directednesss." But once more it does not seem appropriate to talk of a purpose beyond the mere strive for survival. Even accepting that some form of purpose could exist, at least in superior animals, it does not seem to have a determining role in the teleonomic behavior of living beings.

The belief that we live in a world without other goal or destiny that mere being is what led to nihilism. When we speak of a world without purpose we refer to the impossibility of recognizing in the world ends that bear any relation with the human worries and angsts. To describe purpose in the world is to discover a way of being that results compatible with human concerns and that provides motive and orientation to our behavior. To discover purpose, ultimately, is to find a foundation for an ethics. The belief that has consolidated in the last decades, as we saw, is that we live in a Universe that is hospitable to life. Its laws appear fine tuned to make life possible. We see that in the values of the fundamental constants, in the existence of self-organization phenomena that facilitate the appearance of live, as well in emergent structures ever more integrated and complex that present top-down causation. Due to its structure and organization we today know that not only is the Universe hospitable to life, it favors the development of high degrees of mental integration where the mental can be expressed in changes in the behavior. We have also discovered that physical nature described by quantum mechanics can be understood in terms of dispositions and events with a phenomenal internal aspect using therefore concepts close to our conscious experiences. This is just what Russell wanted: "an ultimate scientific account of what goes on in the world, if it were ascertainable, would resemble psychology rather than physics... such an account would not be content to speak, even formally, as though matter, which is a logical fiction, were the ultimate reality"

We could inquire about the reasons that allow humans to act with definite purposes, something that according to all subjective evidence occurs, if they are material entities like any other. Wouldn't any action that obeys a purpose clash with the causal explanations provided by physics? The description resulting from the quantum ontology based on the behaviors of the systems characterized by their dispositions to act and actions determined probabilistically seem to leave room for it. The reader will not miss that, if we accept the unquestionable subjective evidence and some form of physicalism, the notion of matter that one has must be compatible with the appearance of purposeful actions. Common sense indicates that there exists a certain degree of subjectivity in the superior animals that can organize their perceptions. Even when we go down the evolutionary ladder one would expect a certain degree of sensitivity and appetition will be present. The same evolutionary process of the Darwinist vision asserts the existence of primitive forms of interiority and the new physics is not incompatible with them.

If the Universe has a purpose, it should manifest itself in the very laws that favor the emergence of systems with high degree of complexity. These laws not only allow and promote the phenomena required for the generation of life. They also admit a description in which the distinctions between material and spiritual blur. As we saw in the first chapters the regularist position here presented conceives the laws as mere regularities in a world whose nature transcends them. It is a world that ultimately, from a first person perspective it has phenomenal nature, like the one we access from our conscious mind, but it is adjusted to regularities described by the natural sciences, in particular physics.

An interesting exercise within this vision is to invert the famous aphorism of Nietzsche: "Alas, the faith in the dignity and uniqueness of man, in his irreplaceability in the great chain of being, is a thing of the past—he has become an animal, literally and without reservation or qualification, he who was, according to his old faith, almost God... Since Copernicus, man seems to have got himself on an inclined plane —now he is slipping faster and faster away from the center into— what? into nothingness? into a penetrating sense of his nothingness? Very well!"

Maybe that familiarity, that proximity, with the rest of the animal kingdom when seen from the perspective of a Universe hospitable to life and the development of consciousness and thought might be seen from another perspective. Perhaps in a world whose ultimate nature admits at the same time an objective and subjective approach, the notion of purpose could be extended beyond human subjectivity. As we have seen this may not be an exercise devoid of content because on that notion perhaps one can found an ethics.

If the world had no beginning, as for instance the scenario of eternal inflation suggests, it would not make sense to assign purpose to its creation. However, in a world hospitable to life and endowed with an intrinsic nature that has phenomenic character, one can always discover purpose in the beings that inhabit it. It is done by recognizing an orientation towards the development of capabilities that are an ever broader manifestation of the fundamental nature of a world that we have recognized as intrinsically phenomenic and ruled by laws. The emergence of degrees ever more integrated and developed of consciousness would not be but the consequence of the ontological nature of the Universe. The "purpose", not always deliberate, on many occasions resulting form the mere exploration of new possibilities by the inhabitants of the world would be a manifestation, increasingly more complete, of the fundamental nature of our Universe. A Universe that offers constituitively that multiplicity of possibilities of development. That is, a Universe open to creation and exploration by the beings that inhabit it.

Returning to Nietzsche and his aphorism, maybe "the dignity and uniqueness of man, in his irreplaceability in the great chain of being", not only is intrinsic to mankind but to life itself. The process initiated with Copernicus in which progressively mankind lost its singularity guaranteed since antiquity by its privileged relation with God could now invert thanks to the science of the last century in a process of resignification of nature that reestablishes the unity that Nietzsche declared illusory. To exploring this possibility we will devote the last part of this book.

Chapter 16

Monism vs. pluralism: natural religiousness and its historical roots

16.1 Introduction

We have already observed that Nietzsche recognizes as a second form of nihilism the loss of the belief that binds the human beings to a great whole that transcends our personal experience and enables us to devote ourselves to a common welfare. Before analyzing the answers that can be found stemming from the modern scientific advances, we will review in the first part of this chapter two previous instances of religious naturalism where the issue raised by Nietzsche receives a privileged treatment.

The bifurcation generated by Descartes between the physical and mental world led to the development of two opposed currents of thought: the materialist and the idealist. In the first case the problem presented by Nietzsche does not even appear. In the second it appears without appropriate consideration of the advances in the natural sciences. Two naturalist historic conceptions attempt to overcome this dichotomy and deserve special analysis. Both are closely connected with the science of their time. We are talking about Spinoza's conception developed in the second part of the XVII century and Whitehead's conception from the first part of the XX century. The first one is a paradigmatic example of naturalist philosophy where the notion of God plays a central role as the unique substance whose material aspect is nature. The second is the Process Philosophy of Whitehead that takes into account in an indirect way the main scientific discoveries of his time, in particular those of physics: general relativity and quantum mechanics. Whitehead starts with a profound questioning to the lessons usually derived from science and proposes in its place a conception of the world alternative to the metaphysical worldview that developed as a result of the dramatic successes of early modern science that Whitehead calls "scientific materialism". His goal is to develop a metaphysics that he characterizes in the following way: "Speculative philosophy is the endeavour to frame a coherent, logical, necessary system of general ideas in terms of which every element of our experience can be interpreted" (Whitehead 1929/1979). Both attempts are basically unsatisfactory, the first one due to the very poor knowledge of nature that science in Spinoza times provided. Even though the second attempt shows several conceptual advances it is mostly a reaction to the naive naturalistic view provided by the physical sciences and fails to give an explicit connection of the fundamental concepts of the system and the conceptual framework provided by the physical sciences. We will attempt to conclude this book by revising this conception at the light of the recent advances of science and sketching a new naturalistic view.

16.2   Spinoza's conception

Spinoza's thinking had great impact in the second half of the 17th century and the 18th century. He inspired the radical illuminism and was the main questioner of revealed religions and of any divine attributes that the political authorities may have attempted to claim. He also challenged the traditional ideas that had dominated the world till modernity. He renewed the materialistic formulations of the world of Democritus and Lucretius on the basis of the mechanism of Galileo and Descartes. He proposed a naturalist vision, in which nothing exist beyond the natural universe, strongly comprehensive and modern. Starting from a naturalist view of science, he tries to supply a valid alternative to understand the world and the place of mankind in it, without recourse to revealed religion. Spinoza is the model to emulate, given the internal coherence of his system. It combines the results of science of his time with a reverent attitude towards nature, a notion of divinity that satisfies the highest exigences of reason, and his constant ethical pursuit. To reproduce his task has become virtually impossible in our time given the advances of science and of philosophical thought. Science is much more advanced today than in Spinoza's time and therefore it is expected to provide answers in a wider set of areas. This includes many that in his time could only be probed via philosophical elaboration. However we do not have yet a complete unified theory describing all physical phenomena. One can therefore mostly hope for partial successes or insights.

For Spinoza nature is an indivisible, uncaused, substantial whole. Outside of nature there is nothing, and everything that exists is a part of nature and is brought into being by nature with a deterministic necessity. He calls this necessary being, that is at the basis of all things that are governed by its immutable rules, God. Divine providence, so important in the Judeo–Christian tradition is reduced to "nothing but the striving we find both in Nature as a whole and in particular things, tending to preserve and maintain their being." (Curley 1985). Spinoza's thought begins with three central assumptions: it is naturalist, deterministic and rationalist. In this last sense, the influence of Descartes is manifest and is based on his doctrine of innate ideas. According to it, the mind has built into it not only the structure of knowledge but even its content. The Spinozist faith in reason is perhaps the strongest that anyone has ever had. For him, there is a reason for every thing: "For each thing there must be assigned a cause or reason, both for its existence and for its nonexistence." He therefore appears completely convinced of the validity of the principle of sufficient reason. Although its final formulation will come later with Leibniz, the idea goes all the way back to Anaximander of Miletus.

Of these three guiding principles, the first one (naturalism) has ended by being a point of view of increasing predominance. On the other hand, Spinoza's determinism and rationalism are products of its time and have been overtaken by the consolidation of the empirical sciences and quantum physics. Concerning his naturalism, Spinoza claims that all phenomena, in particular human beings, are governed by the same laws. Apparently referring to Descartes, he says: "Most of those who have written about the affects, and men's way of living, seem to treat not of natural things, which follow the common laws of Nature but of things which are outside Nature. Indeed they seem to conceive man in Nature as a dominion within a dominion. For they believe that man

disturbs rather than follows the order of nature." In view of this position he asserts his naturalism: "nothing happens in Nature that can
be attributed to any defect in it, for Nature is always the same, and its
virtue and power of acting are everywhere the same, that is, the laws
and rules of Nature. According to which all things happen..." His understanding is that natural philosophy, which he identified with what
we now call science, had universal application and no area existed that
was beyond its reach. Spinoza therefore discards any possibility of supernatural intervention, leaving no room for the "miraculous" in his
system.

It is clear that naturalism arises as a response to the success of
the sciences that provide a knowledge of reality that is increasingly
broad and unified. This is manifest in the predictions and levels of
control of natural phenomena that modern science has reached, that
have led to remarkable technological developments. Naturalism implies
the denial of dualist cosmovisions that postulate super-natural entities
or miraculous divine interventions. But adopting a naturalist standpoint
implies —and Spinoza was well aware of this—, the impossibility to hold
ethical or religious positions in a noncritical way. Our beliefs have to
be sustainable based on knowledge that we have acquired of the world,
and cannot simply stem from a revealed truth.

A central element of Spinoza's thinking is his monism. It is his answer to Descartes' dualism. The Spinozist system is built as a critique
and supersession of the concept of substance of Descartes. The latter
initiates modern philosophy through the affirmation of rationalism and
the scientific method that arose in the 17th century. The central elements of the Cartesian metaphysics are those of substance, attribute
and mode. The concept of substance is adapted from medieval philosophy and has it origins all the way back in Aristotle. A substance
is something that is capable of having properties or attributes without
being a property or attribute in itself. In addition to this, a substance
is something which does not depend or need of something else for its
existence. It can exist independently of any other thing. For Descartes,
substance is what carries the attributes: "we call the thing which they
[the attributes] are in, a substance" (Cottingham et al. 1985). In his
Principles, Descartes (1644/2010) offers a second characterization: "By
substance we can understand nothing other than a thing which exists in
such a way as to depend on no other thing for its existence." Descartes
believes that there exist two and only two kinds of substances, the men-

tal and the bodily. Each substance has a principal attribute, "one principal property which constitutes its nature and essence, and to which all the other properties are referred" (Descartes 1644/2010). The other properties he calls modes. Descartes also believes that there are only two attributes, each of which is indivisibly linked to one of the substances: thought and extension. Thought is the attribute of the mental, since without thought the mental would be inconceivable. He appears to believe thought is equivalent to conscious content. Extension is the main attribute of physical bodies. "Everything else which can be attributed to body presupposes extension and is merely a mode of extended things" (Descartes 1644/2010). The two fundamental attributes cannot be present in the same substance, "for that would be equivalent to saying that one and the same subject has two different natures." He basically considers that the substances are identifiable through their attributes and that to assign two attributes to the same substance could not be rationally justified.

The problem of how mind and body interact has been one of difficult solution. Descartes himself could not give a satisfactory answer and since then it has been considered a fatal flaw of the Cartesian dualism of substances. Spinoza contradicts the Cartesian viewpoint that substance can only have one attribute and goes on to propose, through a change of the notion of substance, that the latter has a dual aspect: it is extensive and conscious. As long as he conceives the world in terms of extension or in more modern parlance as spatio-temporal, he refers to it as nature and if he thinks of it as conscious he refers to it as God. The Spinozist substance is not something belonging to each particular object, but to the totality of what exists. The particular objects are just modes of this unique and universal substance: "Except God, no substance can exist or be conceived" (Spinoza 1677/2005). There is only one substance and a set of rules that govern everything that exists. There exists nothing contingent since all things are determined "from the necessity of the divine nature to exist and produce an effect in a certain way" (Spinoza 1677/2005). Indeed, Spinoza states that "things could have been produced by God in no other way, and in no other order than they have been produced." Not only everything that happens is determined, in addition he establishes that there is no other possibility. "God (Nature) necessarily is what he is", all appearance of contingency is a byproduct of human foolishness that cannot understand this fundamental truth: "when they see the structure of the human body, they are struck by a foolish wonder, and because they do not know the causes of so great an art, they infer that it is constructed, not by mechanical, but by divine, or supernatural art, and constituted in such a way that one part does not injure another." Nothing can be different from what it is and as a "res extensa" it is no more than a mechanical system that obeys the laws of physics of its time.

In the second part of his *opera magna*, The Ethics (1677/2005), Spinoza establishes the connection between body and mind. He starts from the following statements: "Thought is an attribute of God, and God is a thinking thing... Extension is an attribute of God, and God is an extended thing." He goes on to establish a strict parallelism between both attributes: "the order and connection of ideas is the same that the order and connection of things." He thus introduces the notion that thought and extension are two aspects of the same reality. Although one cannot act on on the other they have behaviors that obey parallel relations of cause and effect. In that sense given the determinism of the physical laws of classical mechanics, Spinozism does not leave any room for liberty. Nevertheless, and in a certain sense paradoxically, since we would have no freedom at all, for Spinoza humans can act in two ways; either driven by passions without understanding the motives of their behavior or understanding those motives to act recognizing their place in the scheme of things: "In so far as the mind understands all things are necessary, so far has it greater power over the effects, or suffers less from them." Spinoza calls beatitude that understanding of our passions "Blessedness is not the reward of virtue, but virtue itself; neither do we rejoice therein, because we control our lusts, but contrariwise, because we rejoice therein, we are able to control our lusts."

Spinoza's conception of God gave rise to the modern version of pantheism that is based on the belief that all of reality is identical with a purely immanent God. The poor physical understanding of the world stemming from classical mechanics led to Spinoza's extraordinarily austere vision of the world, in which humans lack all freedom and basically of any power to affect the ineluctable deployment of the laws of mechanics. This way of life may not be easy, "But all noble things are as difficult as they are rare." In spite of his preoccupation to attain a unified vision of the physical and spiritual world, his system is basically inconsistent. It admits that humans can act driven by ignorance or reason and at the same time it assumes a perfect parallelism between the mental and physical events without any causal interaction between both spheres.

In spite of the apparent detachment from human passions of Spinozist thought and the incoherences we pointed out, it had an enormous political influence that resides mainly in its libertarian vocation that served as a basis for radical illuminism. Spinoza places republican democracy as the best form of government. In democracy " all men remain equal, as they were before in the state of nature" (1670/2007). His democratic ideal establishes as main goal the freedom of thought and speech that Spinoza calls *libertas philosophandi* and it constitutes into one of the currents of thought that led to the French Revolution. Spinoza's vision only has historical value and results from a mechanist and determinist conception of the world that science has left behind long time ago. Nevertheless, the very simplicity and limitations of the scientific knowledge of the time allowed Spinoza's system to achieve a degree of coherence in its naturalism that exceeds that of subsequent attempts.

16.3 Whitehead's conception

Let us recall that Whitehead was educated as a mathematician and logician and he latter became interested in physics —in general relativity and the first ideas of the new quantum mechanics— and philosophy. Whitehead's thought is generally considered to be among the most difficult to understand in all of the western tradition. Even professional philosophers struggled to follow Whitehead's writings. Whitehead rejects the idea of separate and unchanging bits of matter as the most basic building blocks of reality, in favor of the idea of reality as interrelated events in process. The resulting vision as it was presented in the original works results extremely abstract and at first sight very removed from the scientific notions we have discussed in this book. However, its concepts can be useful in an analysis more connected to scientific developments and we will use some of them in the following sections.

In his philosophical master work Process and Reality (1929/1979) he developed an ontology based on a new notion of actual entity or actual occasion of experience. This notion substitutes the Aristotelian notion of substance or Leibnitz notion of monads as the foundational elements of reality. He called the actual entities found in nature actual occasions of experience. Each occasion of experience is "atomic" in the sense that they are not composed by other actual entities and has a spatial and temporal dimension, is causally influenced by previous occasions and causally influences future occasions. Each actual entity has causal power over others: "to be is to create". Each occasion consists in a process of "prehending" other occasions and reacting to them. Without the existence of a total identity between the actual occasions and what we have called events, there exist strong similarities, for example in what concerns their properties that Whitehead considers as actualizations of what he calls "eternal objects" in the actual occasion considered. The "eternal objects" are identified, when abstracted, from the actual entities. Thus the color red transcends the particular occasions of its occurrence.

In Whitehead's philosophy, every actual entity has a physical and a mental pole. Whitehead speaks of actual entities as "dipolar". The physical and mental poles are aspects of every actual entity. For Whitehead, human consciousness is a higher form of mentality but not the only form. Whitehead conceives God as an actual entity. But an actual entity that plays a fundamental role: God is the originator of the contingent world. He says "God is the ultimate limitation, and His existence is the ultimate irrationality... He is the ground for concrete actuality" (1925/1997). Thus, God is at the origin of choices among possible worlds. We only can know which world has been chosen through empirical evidence. Thus the "ultimate irrationality". One cannot, through the sole use of reason, determine how the contingent and empirical reality must be. Nevertheless, Whitehead denies the existence of a creator God that precedes the Universe. He rather argues that the choices that determine the contingent Universe result from acts of the actual individual entities influenced by God.

In his review on Whiteheads philosophy, Donald Viney (2014) recognizes two complementary natures in Whitehead's conception of God. One primordial and one consequent: "The primordial nature is God's envisagement of all possibilities; in the idiom of Leibniz, it is God's knowledge of all possible worlds. It is called 'primordial' because it represents what could be in a sense not tethered to the actual course of events. It is logical space, deficient in actuality apart from the consequent nature says Whitehead. ...The consequent nature is the record of all achieved fact, a perfect memory of what has been. Whitehead speaks of the 'objective immortality' of the world in God. The two natures work in concert in the process of God's interaction with the creatures. ...The deity receives the world of actual occasions into its experience; then, comparing what has actually occurred with the realm of pure possibility, God informs the world with new ideals (new aims),... It is God's relevance for the world as a "lure for feeling," urging the creatures to strive for whatever perfection of which they are capable."

Whitehead's conception of God is many times included in what is called Panentheism, which considers that God and the world are not the same thing but that God is involved in every event of the world even if he transcends it. The contingency that we discover in nature leads to consider that God cannot be immutable because God's knowledge of contingent events. For Whitehead an essential attribute of God is to be involved in temporal processes. And he therefore has a certain degree of contingency: God affects and is affected by the world. In this point the panentheist vision, called many times *Process Theism* departs from traditional theism and Spinoza's pantheism for whom the world obeys laws too strict and God is immutable and eternal. Whitehead's scheme, based on "actual occasions" limited in space and time seems to lean towards a pluralist vision of fundamental objects. Nevertheless, Whitehead resisted pluralism, maybe due to his reluctance to assign a separate actuality to God, which would ultimately be the creator of the world: "There are not two actual entities, the creativity and the creature. There is only one entity which is the self-creating creature."

Whitehead's philosophy, although inspired in the discoveries of physics of his time is not strictly speaking naturalist. Indeed, as we observed it is a speculative conception that is stated in terms of concepts that do not have a direct connection with the scientific ones and therefore with the ones that have univocal meanings of formal character and yield the best descriptions of what we know about nature. He rather starts from a justified questioning of the language of scientific materialism, but he substitutes it with concepts without direct reference to the concepts that science —in particular physics— uses in its theoretical constructions. For example, the notion of *actual occasion of experience* combines without discriminating among them different notions that appear in the axioms of quantum mechanics. Among others those of state, event and Hamiltonian evolution, which contributes to the obscurity of the concept.

16.4   Quantum physics and the ontological pluralism of actualities

The worldview that derives from quantum physics does not, prima facie, appear to be compatible with the Spinozist monism. In the latter, God and nature are identified. Nature is an indivisible, uncaused, substantial whole. Outside of nature there is nothing. Everything that exists is part of nature and is brought into being by nature with a deterministic necessity. Spinoza calls Nature or God, depending on the attribute through which it manifests itself, to this necessary being that is at the basis of all things that are ruled by his immutable rules.

On the other hand science identifies multiple entities, each of which, though not necessarily independent, is at least distinguishable from the others. This position, usually known as pluralism, has a long tradition in British philosophy from Occam to Russell. We have identified two types of fundamental entities: the individual objects, that is the systems in given states and the events. The states are non-local entities that can suffer sudden changes and yield new states when events take place. We have characterized the states by their disposition to produce events. In that sense they have an existence characterized by their potentialities. Their individuality, as we shall see, is always approximate and conventional (since in reality all systems are in interaction and entangled with the environment). Concrete reality is constituted by events localized in space-time. As Whitehead recognized: "the event is the ultimate unit of natural occurrence." The objects that evolve, including living beings, manifest themselves as processes involving a large number of fundamental systems in definite states in succession and coexistence. The events, once produced, are definitive: they are produced in a definite region of space and time. In that sense relativistic physics, which refers to events, establishes a strict causal order associated to the nature of space-time itself. True or false statements will refer to events, and since the latter are objective, they become the building blocks of a definitive past. In the words of Peirce "the past is the sum of accomplished facts" (Buchler 1955).

Seth Lloyd has acutely observed that the quantum processes of actualization of events are basically processes of creation of information. According to Lloyd (2010): "Let us look more closely at how quantum mechanics injects information into the universe. The laws of quantum mechanics are largely deterministic: most of the time, each state gives rise to one, and only one, state at a later time... Every now and then, however, an element of chance is injected into quantum evolution: when this happens a state can give rise probabilistically to several different possible states at a later time." There exists a mathematical measure of how much creation takes place at each actualization in events, introduced by Claude Shannon. The idea is that one can measure the ignorance about a system that has different alternatives of behavior and one does not know which of them will be chosen. Such ignorance is called Shannon entropy. When the event takes place ignorance disappears. The information created by the process is by definition equal to the Shannon entropy that was present before the process. Originally, this concept of information was introduced to measure the quantity of ignorance that can be eliminated when one receives a message in a process of communication between an emitter and a receiver. In this context it is also called classical information. A bit of information is the quantity received when one communicates the result of flipping a coin and one gets the message: heads or tails. A computer is basically an information processing system where some input information is processed, stored and eventually ends up producing an output.

Purely classical processes like the collision of two macroscopic masses may be considered as examples of information processing. The same way we characterize ignorance in classical systems we do in quantum ones, but referring to the outcome of a quantum measurement. The passage from potentialities to concrete events injects information into the universe. Although the evolution of the potentialities is deterministic, ruled by Schrödinger's equation, a random element is introduced every time an event takes place. When this occurs a state may give rise probabilistically to several different possible outcomes. Quantum systems allow to explore a wide spectrum of possibilities. In quantum systems the number of alternatives for a composite system grows exponentially with the number of components. When a quantum choice occurs and a event is produced, new information is introduced into the world.

"Every detail that we see around us, every vein on a leaf, every whirl on a fingerprint, every star in the sky, can be traced back to some bit that quantum mechanics created" (Lloyd 2010). In the moment prior to the production of an event, we have a state of a complex system composed of microscopic systems, the environment and eventually measuring devices. It has arisen through reduction processes of previous states and through the Schrödinger evolution of the interaction of its parts. The event that this state is about to produce therefore results from a conditioned election of a great number of previous events. In process philosophy (Whitehead 1925/1997) the notion of prehension is used to describe the synthesis of previous influences leading to the production of a definite event. In metaphysical terms, it is a creation process and also an "experience" where "many past events function as data. ...a 'synthesis', a new single reality [a new event] partially constituted by many past realities" (Hartshorne 1979). It happens when an event takes place. Quantum physics therefore leads to an *ontological pluralism of actual entities*, the events. The pluralism of actualities here presented conceives the universe as made up of many entities, each of which is definitely distinguishable from every other. Recall that the notion of event and property is related in quantum mechanics to mathematical objects called projectors. Events are instantiated properties. Thus, the traditional notion of substance does not seem applicable to these entities because as we have already discussed, the notion of substance refers to something that carries properties but does not reduce to them.

The notion of prehension introduced by Whitehead seems to account for this process of creation. For Whitehead (1925/1997), "An event has to do with all that there is, and in particular with all other events... There is thus an intrinsic and an extrinsic reality of an event, namely, the event as in its own prehension, and the event as in the prehension of other events" Even though Whitehead admits that events have an internal aspect, as we have emphasized in previous chapters, his conception ignores, again following the empiricist tradition, the existence of states. Whitehead refers to the effects that an event has on another, ignoring that events are mediated by the states. We on the other hand, claim that only through the states is the concatenation of events ruled by laws.

16.5   Ontological monism of potentialities

We will see in this section that modern physics leads to a form of monism that shows that there is in fact a natural binding of human beings among themselves and with the rest of the Universe, and therefore this unity that Camus desired and Nietszche considered unattainable is in fact at the core of the physical reality when conceived in terms of states and events.

Individual objects were defined as systems in states. If one were to establish a hierarchy between the fundamental concepts, states and events should be considered as more fundamental than systems and properties. In quantum mechanics one speaks of properties belonging to events. The independence or not of an individual is defined by its state. For instance, a neutron will be an independent individual as long as it is in a non-entangled state. In this case all its possible behaviors will be determined, including the probabilities for producing events. On the other hand, as a part of a two-neutron entangled state, each neutron will not be an independent individual but what we call an individual fragment. Recall the words that Asher Peres assigned to Sagredo when in Chapter 7 we discussed the issue of entanglement of two polarized photons: "Sagredo. There is a paradox only because you force on this physical system a description with two separate photons. These photons exist only in your imagination. The only thing you have really prepared is a pair of photons, in a ...[definite total] spin state. The pair is a single, indivisible, nonlocal object...". A system which is a component of an entangled state, will never be an independent individual because its state will not be enough to describe all of its possible behaviors. Independent individuals are ideal entities very well approximated by certain microscopic objects: for instance a neutral spinning particle in a vacuum chamber after going through a Stern-Gerlach device that is in a state $|z; up\rangle$ is very approximately an independent individual. Macroscopic objects, like measuring devices or ordinary objects as a table or a book, are always entangled with the environment, and therefore are individual fragments.

In the real world, the effects induced by the entanglement of a macroscopic system with the environment are extraordinarily strong and quick. If one attempts to prepare a system in a superposed quantum state for an object like a macroscopic needle of a measuring device interacting with photons, the interaction will lead to a complete loss of coherence in the superposition within fractions of a second. This process is called decoherence. Decoherence is an extremely fast process for macroscopic objects, since these are interacting with an environment composed of many microscopic objects, constituting an enormous number of degrees of freedom. Moreover the decoherence time decreases exponentially with the number of degrees of freedom present in the environment. For instance, for a pendulum of mass 1 gram, with period

one second and damping time of one minute, if one were to place it
in a quantum state that is a superposition of the pendulum at two
different positions, separated just one micron, the decoherence time is
$10^{-16}s$. The measuring device loses its individuality when it interacts
with the environment. It becomes an entangled system with it and a lot
of its physical properties will be lost due to the non-local correlations
of the complete system. As we have mentioned, in an entangled system
the whole is more than the parts. There should therefore be no doubt
that the information initially present in the measurement device gets
smeared through the total system including the environment. In order
to fix ideas let us consider one of the particles with spin in an entangled
pair. It is an individual fragment. It still makes sense to speak about
the state of this fragment. The state of one of the isolated particles with
spin would be a Von Neumann density matrix such that with probability 1/2 its spin is down and with probability 1/2 up. The state of the
fragment is not enough to determine its complete behavior because of
the existence of top-down causation in entangled systems. For instance,
the correlations between the events produced by particle 1 and 2 are
not explained by the state of particle 1. Each particle of the set is in
this case an individual fragment and is related with the other in the
sense that it shows behaviors that are not explainable only in terms of
its state.

While event pluralism is the most distinctive characteristic and manifestation of the multiple nature of apparent reality, individual objects'
pluralism has a meaning more conventional than fundamental, due to
the non-local and holistic character of the states. Indeed, if we consider
a fragment, the knowledge of the state of the individual fragment is not
enough to predict its disposition to produce events. On the other hand,
perfectly independent individuals do not exist. To begin with, as we
have emphasized, states have non-local aspects. For instance, the state
of a free particle leads to a probability not strictly vanishing of finding
the particle at any point of space. Whereas in a classical system of N
identical particles one can identify each particle through the position
it occupies in space, in a quantum system this is not possible because
particles do not occupy definite positions in quantum mechanics and
one always has a non vanishing probability of finding particle (1) in the
region assigned to particle (2). Only to the extent that we can isolate
completely one of the particles, for instance placing them in a closed
container or magnetically confining them, we can treat them ignoring

the others. Even in those cases the particle interacts with the walls of the container or with other particles of the environment since a perfect vacuum does not exist. In the case of macroscopic systems one needs to add the ease with which they get entangled with the environment. Summarizing, individuals are conceptualizations convenient for the analysis of the behavior of things. They allow to predict a good fraction of their behaviors, though they only have an empirical value behind which a holistic reality hides, that ultimately refers to the universe and its state. The notion of isolated system is conventional, similarly to that of independent state. All systems interact with others and their state is always (at least partially) entangled with the rest of the universe. The notion of isolated system is as arbitrary as the humorous characterization of physicists discussing the existence of a "spherical cow of negligible mass". We must, nevertheless, refer to it because physicists need to have everything under control in order to verify the laws with absolute precision. That is why in many occasions we consider as perfect a vacuum that can be created in the lab, at very low temperatures or vibration isolation, as for instance certain in gravitational wave detectors. But ultimately one can consider that the only isolated system is the whole universe. We had recognized in individuals similarities with the traditional notion of substance. But unlike the notion of substance as undifferentiated substratum, individuals have characteristics that allow to single them out as a system with a set of identifiable behaviors, that is, in a recognizable dispositional state. This therefore elucidates the notion of substance, making it concrete and empirically analyzable. By not being an undifferentiated whole and becoming analyzable one can discuss the behavior of a particular thing and its state. This in spite of the fact that it cannot be isolated from the rest of the universe without losing some of the information that determines its behavior. An element of key importance that was introduced when the emergence of new levels of behavior was analyzed is that, as long as individuals behave like fragments of more general systems, they exhibit downward causation. That implies that the behavior of the individual in question cannot be completely understood if one does not think of it as part of a bigger system.

From the quantum ontology a pluralism of actualities has arisen: the events, whose existence is contingent and define the concrete reality. Events are ordered in space-time creating a causal network. *The tendencies to produce events are determined by the state of the fragment and*

the rest of the Universe. The latter can be identified as the unique substance from which all individuals and their choices manifested in events
emerge. In other words, it is a pluralism of contingent events produced
by fragments of a unique substance. Physics as an empirical science has
led us from the facts expressed in events to the identification of systems
and states. From them we have recognized the individual objects as
the substantive aspect of the quantum ontology and finally observed
that they must be identified as fragments of a unique substance. Events
generically occur in complex individuals composed by many fragments
that, when an event is produced, take part in the choice. In the typical
situation studied in quantum physics, the individual consists of a micro
system, a measuring device and an environment. Together they produce
the event. Their behavior is determined by what we called in the chapter about emergence as top-down causation. Quantum mechanics allows
to assign precise probabilities to measurements made on systems whose
state has been previously prepared. In this case one has the possibility
of identifying in an unambiguous way the individuals that take part in
the production of the event. In our quantum scheme, the fragments
where choices take place are identifiable, and can be considered independent (in the sense of being capable of determining the event). The
choices are made from within different alternatives, and the events that
do get actualized are therefore contingent. The unique substance is the
state of the universe. It gives a complete and ultimate characterization
of the dispositions to produce the various events, leaving the choice in
the sphere of the individual fragments.

In a scheme in which one ultimately ends up speaking of a unique
substance that could be identified as the state of the Universe, the individuation of bodies must be explained. Could the notion of individual fragments that we have developed up to now and the pluralism of
events help to understand the processes of individuation such as those
that take place in a quartz crystal? We would like to investigate the
role that individuals play in the processes of configurations of identifiable parts as the one mentioned above. What identifies a thing as such?
Let us think of a piece of quartz crystal. We could start by observing
that it behaves like an approximately rigid body and therefore its parts
have distance relations that remain unchanged and it offers resistance
to be deformed. It behaves as a unity, it has a well defined surface that
encompasses it, it is quite hard and can etch steel, and at microscopic
level it has a certain crystalline structure (trigonal rhombohedral) and

it is composed by silicon oxide. From a physical point of view it must be considered a fragment individual. Its behaviors, for instance its color and transparency, depend on the light illuminating it and its behavior is typical of a classical object that results from its interaction with the environment and decoherence. It nevertheless has an identity resulting from the characteristics we have listed that present a set of typical behaviors, that express themselves in the multitude of events that define its appearance. The particular individual fragments that at each instant represent the piece of quartz changes constantly due to the production of new events when the crystal interacts with the environment and the eventual modifications it may induce in the latter. Recall that after the production of an event the state that represents the fragment changes and therefore the crystal is not always associated with the same individual fragment. The diverse individuals that represent the quartz at different times have common elements that allow to recognize it as an object that persists in time. They manifest themselves in the systems that compose the crystal, the atoms of silicon and oxygen and certain characteristics of the states like those that determine its crystalline structure. The states change but assign similar probabilities to find the atomic nuclei in configurations that reproduce the crystalline structure. A given piece of quartz crystal is therefore associated with a set of individuals that change in a permanent transformation process. They produce new events at each instant but they maintain certain characteristics in the nature of the system that compose them and in the dispositions they present while being organized in certain structures like the crystalline ones. In the crystals, the chemical composition and the atoms strongly bound into a crystalline structure that compose it remain invariant and the mentioned changes take place in the states of the system and the events that they lead to. The fragment individual associated with the piece of quartz has an eminently ephemeral nature in time although it has precise spatial limits. The permanence of objects is only approximate, objects continue to exist even when they cannot be seen, heard or touched, but we recognize them as the same object due to our great ability to recognize patterns and similarities, since they are never exactly identical to the last time we encountered them.

Summarizing, for Spinoza everything that exists reduces to a single substance and its modes and attributes. This type of monism is usually known as existence monism. In our perspective physics, as an empirical science, has led us from the facts expressed in events to the identification of systems and states. From them we have recognized the individual objects as the substantive aspect of the quantum ontology and finally observed that they must be identified as fragments of a unique substance. Since the states characterize the disposition of a physical system to behave in a certain way, that is, they manifest their potentialities of action, the thesis here presented of a unique substance, can be denominated a *Monism of Potentialities*.

It is curious that one of the most universal and impersonal of sciences reveals to us that behind the apparent diversity there exists and omnipresent entanglement among objects that constitute the universe. They are part of a unity without which it is not possible to understand their behavior. If to this observation we add the one made about the emergence of wholes that are superior to the sum of their parts with downward causation every time the quantum entanglement that connects us to the whole universe is present, it leads us to question: how do these properties manifest themselves in each individual object? Especially in humans: they may manifest themselves as religious feelings, ethical imperatives, artistic inspiration? Without reasons to believe in a world without purpose or unity, thanks to the change in the conception of the world provided by contemporary science, the third form of nihilism —stemming from the failure of thought in the search of giving meaning to the world— does not even arise.

Chapter 17

Emotions, ethics and free will

17.1 Introduction

In the previous chapter we noted that the constituents of the Universe, the systems in certain states, are part of a unity without which it is not possible to understand their behavior. The question therefore arises of how do the states manifest themselves when we access it directly, that is, in our conscience. Let us recall that like Whitehead we assume that "There is thus an intrinsic and extrinsic reality of an event". The latter are the events that form part of the objective reality accessible to our senses, the former are the events in our brains that manifest themselves, among other possibilities, as sensations. Whitehead's vision, stemming from the empiricist tradition, started with events as basic components of its ontology. In our case the ontology is formulated in terms of events and individual objects: that is systems in certain states. Here we will therefore concern ourselves with analyzing: which is the intrinsic form of the states whose extrinsic characteristics are described by quantum physics?

17.2   Emotions as the internal aspect of states

The thesis of this section is that moods, emotions and desires are not only states of the mind but they may be also characterized in terms of one of the fundamental elements of the quantum ontology. In the same way that the sensations are internal aspects of the events in our brain, the moods and emotions may be considered as internal aspects of the dispositional states in our brain. They would therefore be an intrinsic part of human —and at least superior animals— actions, they inform us about the states of the conscious subject and of how they orient its behavior.

Both moods and emotions have certain common aspect: they manifest themselves in modifications in the way we relate to the world. A mood, for instance a depression, and an emotion, for instance indignation in the face of injustice, differ in that the second has an identifiable origin related to certain occurrences, whereas the first may not. In the case of indignation, for instance, something that someone else did that we do not approve of. There does not exist a clear cut distinction between emotions and moods. An emotion of sadness occasioned by a concrete situation can turn into a depression. A specific fear, for instance being in an unprotected house in a dangerous neighborhood, can derive in a generalized state of anxiety and paranoia.

Emotions very likely have a quite ancient origin and go back evolutionarily at least to superior animals like mammals. At least this is the case for a set of emotions one can consider primaries: happiness, sadness, fear, anger, surprise and dissatisfaction. All refer to internal states of the individual and their harmony or lack thereof with the environment and tend to provide the individual with strategies of survival.

Emotions start with an appraisal of a situation that gives rise to behaviors that generically can be characterized as of attraction or repulsion, depending on if the situation is judged as beneficial or harmful. The observation indicates that an individual will judge about the positive or negative character of the occurrence that moves him according to criteria that not always are limited to his personal objectives.

A question that has been in the middle of the discussions about the role of emotions, concerning their impact on the rationality of our actions: are emotions a help or a hindrance to find the best answers?

A second question is about the role of emotions in moral questions. Concerning the first issue, if we understand by rational behavior the one which shows coherence between our beliefs and the evidence that exists for them and the adjustment of our actions to the motives for action, it is not clear that such reason-oriented action is even well defined. On the one hand it is not clear how deliberation can succeed in leading to action. Indeed, it is not evident what to assume about the reasoning processes of the agents that participate in a given matter if there is not some motivation that precedes the deliberation. On the other hand, the number of objectives that can be pursued and the number of strategies that can be devised are usually extraordinarily large. A rational decision should be based on an analysis of all the consequences that follow each of the alternatives. In artificial intelligence and cognitive sciences it is assumed that a drastic pre-selection of the possibilities to be analyzed should take place. This is known as the frame problem.

This problem does not seem to arise in human beings. At least not with the same degree of acuteness. Emotions seem to play a natural role of defining a frame. As we have seen, they guide our attention to the facts directly involved and lead us to consider only a small number of alternatives of action. The way in which this selection operates is not well understood although in certain cases one can conjecture about some of its mechanisms of action. For instance, if someone hates another person, it will assign intentions of action based on a distorted selection of previous observed behaviors. If someone loves another person, the desires and objectives of the loved individual will be taken into account when making decisions. It is obvious that these drastic simplifications in the possible scenarios that emotions generate lead to errors. In terms of the quantum ontology of events and states, emotions alter the set of probable behaviors that the states assign, increasing some probabilities and decreasing others. This leads to a simplification in the decision processes that lead to our actions. Emotions that will play an important role later, as pleasure and pain lead to frames that attempt to repeat the pleasurable situations and avoid the painful ones.

In what concerns the role of emotions in moral behavior, "virtuous" behaviors like those associated with empathy and sympathy have been observed in a large number non-human species. Even Darwin (2010) noted that certain forms of morality are a natural tendency. Preston and de Waal (2002) in their important work on the communication of emotions in animals have observed that empathy, defined by a "shared

emotional response between an observer and a stimulus person.', Fesh-
bach (1975) and sympathy, understood as the capacity of an observer to
feel sorrow for another that is suffering, are common in higher mammals.
There exist good reasons why such emotions have developed evolutionarily, for example, among the primates. Indeed, the latter form complex
societies; they form couples in search of food, protection or status, they
establish alliances to protect themselves. That requires the ability to
recognize facial expressions, voices or other forms of expression and to
respond quickly to them. Empathic behaviors very similar to those
of humans have been observed. For example: "Kidogo, a twenty-one
year old bonobo [Pan paniscus] at the Milwaukee County Zoo suffers
from a serious heart condition. He is feeble, lacking the normal stamina
and self-confidence of a grown male. When first moved to Milwaukee
Zoo, the keepers' shifting commands in the unfamiliar building thor-
oughly confused him. He failed to understand where to go when people
urged him to move from one place to another. Other apes in the group
would step in, however. They would approach Kidogo, take him by the
hand, and lead him in the right direction. Care-taker and animal trainer
Barbara Bell observed many instances of spontaneous assistance, and
learned to call upon other bonobos to move Kidogo. If lost, Kidogo
would utter distress calls, whereupon others would calm him down, or
act as his guide. One of his main helpers was the highest ranking male,
Lody. These observations of bonobo males walking hand-in-hand dispel
the notion that they are unsupportive of each other" (de Waal (1997)).

In what concerns the relation between emotions and moral issues.
Far from sharing the Socratic or Spinozist position that virtue resides
in knowledge, or to consider like the stoics that emotions are no more
than irrational beliefs, we hold like Aristotle that the question resides in
learning to feel the right emotions in the right circumstances. For Aristotle virtue consists in achieving a combination of abilities, rational,
emotional and social. Virtue, when talking about ethical behavior, does
not consist in acting according to rules, it is acquired in practice looking for a suitable combination of rational and emotional aptitudes. In
certain occasions, only emotions seems to arrive at the morally correct
reaction, as follows, for instance, from studies about altruistic behavior
of those that participated in the rescue of Jews in World War II.

Kristen Monroe (1996) did interesting field work in which she in-
terviewed rescuers of Jews during World War II about the origins and
motivations of altruistic behaviors. That is, actions carried out with

the conscious objective of helping another person without any expectation of reward for whom carries out the actions. The altruistic act can also imply for the person carrying it out putting at risk their well being or other personal objectives. After many interviews, Monroe identifies as the central characteristic of the altruist a strong bond with others manifested in a "shared humanity" and recognizes two systematic behaviors in altruistic acts. The first is their spontaneity, the acts do not appear to result from a rational prior analysis. The decision to risk their lives appeared spontaneously responding, according to them, just like any other person in the same situation would have done. The second characteristic is that these people seemed to obey a irresistible sense of responsibility. The typical response was "what else was I supposed to do?". We therefore see that altruistic behavior does not appear to result from a rational analysis nor from a deliberation, it is rather the consequence of a feeling of compassion or responsibility. If emotions are related with our dispositional states and these states in most cases not only involve the individual but includes top down effect from the environment, it is not so surprising that many moral behaviors have an emotional origin.

17.3   Freedom of the will versus random choices.

Each living being has as a primary aim his striving for survival. Given the extraordinary quantity and variation of beings, each one propelled by the same basic purpose, the competition is intrinsic to the process. The purpose of some only can be realized at the expense of others. Each living being pursues in the first place a selfish goal that is to continue being and is faced before anything else with the risk of ceasing to be. Already in the superior animals we certainly see behaviors that show that other goals, for instance those that involve protecting their progeny or solidarious behaviors as we mentioned previously.

In humans, the self-preservation impulse is not blindly followed, it becomes one among many motives that define their deliberate acts. It is in the context of a subject that feels free that appears the feeling of duty, the notion of responsibility and the possibility of classifying our actions as good or evil.

Most of us are convinced of our ability to choose between different possible courses of action. Without freely chosen actions, it becomes difficult to assign responsibilities, justify praise or to blame a person. We can understand free will as the ability to make choices about matters that are not determined by past events. In a deterministic context it appears difficulty that more than an outcome could be possible given that the future is completely determined by past events and as a consequence it does not appear possible to justify the existence of free will. Although there exist points of view, known as compatibilists, that consider that free will is compatible with determinism, in the majority of cases they need to modify the concept of freedom of action. For instance, by associating the concept to that of making decisions on the basis of rational arguments. In agreement with the incomaptibilists, we believe that in a deterministic world, free will is an illusion.

Quantum choices are not ruled by any deterministic law, they introduce an element of randomness in the process of decision. However, purely probabilistic randomness does not seem enough to ensure freedom of the will and the possibility of responsible acts. It is only in the context of what we called regularist physicalism that we are able to recognize the elements required in a free act. In fact, this kind of physicalism allows the possibility of causal openness in nondeterministic processes. In other words there could exist origins for our action that are not regulated by any physical law beyond the probabilistic restrictions that they impose . Actions would occur as a consequence of two elements. The dispositional state that define the intention to produce the effect and the precise choice of one of the possible effects.

Let us recall the three basic hypotheses which we have emphasized repeatedly: a) *Regularism.* Physical Laws describe the regularities exhibited by the material processes, whose existence precedes them. Theories are therefore under-determined by our experience, although in practice physicists are unaware of this under-determination. b) *Ontology of objects and events.* The mathematical formalism of the physical laws describes a world of events with certain properties and systems in certain states, which we have dubbed objects. In particular they present phenomenic aspects like sensations or emotions. c) *Indeterminism.* The quantum ontology is indeterministic and randomness is deeply ingrained in its principles.

The thesis of this book is that these three hypothesis allow us to explain the causal efficiency of free choices. We claim here -as we have done throughout the book- that physical laws do not exhaust reality,

they simply describe its regularities. The fact that the quantum dispositions lead to probabilistic predictions does not imply that the action
is random in the sense that it would be if we flip a coin or run some
other random process prior to the decision. Let us recall that what the
quantum theory says is that the passage from the world of potentialities
described by the states to the actuality of the events, that is, the process
of actualization of events, is not ruled by the laws of physics beyond the
probabilistic predictions. An individual can choose, without implying a
deviation from the rules of chance. The restrictions imposed by a regularity that is probabilistic in nature can be illustrated in the following
example. In the English language each letter of the alphabet appears
with a certain probability. That does not get in the way of writing the
sports section of a newspaper or Shakespeare writing a tragedy. The
frequency with which certain letters appear may certainly vary from
one text to another just like any finite sequence of numbers may appear
in a random sequence.

A possible physical implementation of this freedom beyond the randomness of the probabilistic rule could be based on the observed monism
of potentialities. As we observed as any macroscopic system, we are
strongly entangled with the environment. Our choices should obey the
laws of quantum mechanics and therefore their probability will be determined by the entangled state with the rest of the world. A possible
additional criterion that is physically implementable and at the same
time could have implications on further choices of the same individual
could be to link the choice between with the preservation or not of its
degree of entanglement with the rest of the world [1]. If the degree of entanglement would be related with the degree of harmony of our actions
with the rest of the humans and nature, we would have a simple way to
allow morally significant choices with quantum physics.

From this example stems a possible implementation of free will. An
answer that perhaps is novel to the meta-ethical problem that deals
with the foundations of our ethical statements. Let us assume that our
freedom is reduced to choosing among a finite set of options that are
in the end of the same kind: as in the example to act in harmony with
nature and the other, defined precisely in terms of entanglement; or to

---

[1] A way of physical implementation of this kind of possibility is choosing what
kind of quantum observables will be actualized. Those whose outcomes depend on
the complete system including the environment or those that only depend of the
system. This kind of measurements where analyzed in Gambini *et al.* (2010/2011).

act against that option attempting to isolate ourselves from external influences an concern ourselves with our own interests. The proposal is that the options have objective character and are always the same. The choice between these options would apply universally, that is to all people regardless of culture, religion or other distinguishing feature. Thus one would say that a fundamental moral standard results from this distinction consisting in considering as good and desirable the behaviors which are more prone to the overcoming of the fragmentary aspects of our existence, by following dispositions that result from our openness to the rest of the world. Any moral choice in favor to the this option presupposes as we will do based the analysis of the previous chapters that the behaviors that favor this choice are tuned with the aims and purposes of the state of the Universe. The progressive conviction that such goals involve laws and phenomena amazingly oriented to the development of higher forms of life and consciousness favors the adoption of this position starting from strictly naturalist foundations.

Humans are therefore responsible for their actions and omissions. Such responsibility is intimately linked to the possibility of determining by ourselves our behavior and our ends and purposes. Having free will is ultimately to be the masters of our own lives. Maybe the fundamental freedom of mankind does not reside in being able to act in different ways in a given situation but in using to the maximum their abilities heeding a call we are all exposed to as human beings. It is a call to the responsibility that builds us up from our actions in a process of development that is purely creative. In this sense, freedom would not reside in choosing among options but in identifying possible options obeying a calling that due to its universality, offers infinite possibilities of action.

If any acts may be in the end creative, what constitutes goodness, when is an act virtuous? Up to the extent that the monism of potentialities indicates that we are nothing but individual fragments, to do what benefits our nature does not necessarily lead to an egotistical behavior. One could assume that we have the possibility to act following the most general dispositional state that as we saw, is the state of the universe that acts upon us through downward causation. As in entangled states, parts can have behaviors that stem from the state of the whole. In that sense, up to the extent that our behavior obeys the deepest nature of the individuals it does not necessarily have as goal its self-interest, it can respond to a broader interest. Altruistic behaviors are not foreign to humans, but manifestations of its deepest nature. Given our capability for self-determination we have the possibility of progressively insulate ourselves of our sensitivity to the needs of others or open ourselves to it. This is the fundamental ethical distinction, even if it does not derive from any pre-established norm like an eternal truth of some conception of an ideal form of good.

Humans have ethical behaviors that in good measure depend on the history of the development of our species. Ethics develops historically as science in the sense that a greater understanding of societies and a deepening of the behaviors opened to downward causation can contribute to its development. This latter aspect would be of universal nature and common to all intelligent species conceivable, since it is related to the nature of any individual. In that sense, irrespective of the contingent differences of each evolutionary development, there would be a common source of morality. Morality expressed in this way would not manifest itself in a Kantian imperative nor in a sense of duty but it would be expressed in a feeling of sympathy to others so intense and persuasive that it would leave us in pain and guilty if we did not follow it.

Chapter 18

Creativity and God

18.1 Creativity

Within the context of the monism of potentialities, the basic elements that constitute fragments of the unique substance of the physical reality are what we called individual objects. Recall that they are states of a certain systems like a molecule or a crystal. The unique substance as we have already advanced is the state of the whole Universe that acts by top-down causation over all the individual fragments. If anything deserves the name of God within this conception would be this.

Each of these individuals is characterized for its disposition to produce events —given by its quantum state– and their freedom to choose which event to produce with the sole limitation of the probabilities derived from their disposition. Individuals are actualized in the acts that lead to the events that constitute the world. Recall that Lloyd (2010) summarized this process as follows, "Every detail that we see around us, every vein on a leaf, every whorl on a fingerprint, every star in the sky, can be traced back to some bit that quantum mechanics created." Every individual is by nature an entity disposed to create an event or to participate as part of a wider system in a process of creation.

The processes of creation of events occur in succession in the same entity, for instance in the same cell. After each actualization the state of the entity changes abruptly in the reduction process. In that case each election contributes to determine the following state of the same entity, who therefore has a certain ability of self-determination. Each

individual entity is therefore, up to a certain extent, self-creative, also
known as causa-sui: it creates in the world new events and it creates
itself by determining the state into which it will transform after its
choice. Creation and synthesis of all the elements that determine the
state prior to the event and freedom and originality, as long as it gives
rise to new events and states, summarize the content of the actualization
processes of the individuals. Creativity seems so difficult to understand
because it involves the appearance of new patterns of events that have
never existed before. In a classical deterministic universe novelty can
only result from the development of preexisting causes. In this kind
of world, a new entity, a new idea or a new form of life is nothing
but the manifestation of a preexisting and determined possibility. This
form of understanding creativity is originated in Platonic thought. "The
ancients, Platonists to a grater or lesser degree... imagined that being
was given once and for all, complete and perfect, in the immutable
system of ideas; the world which unfolds before our eyes therefore add
nothing to it..." (Bergson 1907/2005)

In a quantum universe creativity plays a crucial role and new forms
of organization emerge. Natural selection is in line with this conception
of creativity. For Darwin, creativity is not beyond nature. The latter
itself gives rise to the infinite forms of life. It is therefore not surprising
that the modern notions of creativity are strongly inspired in the theory
of evolution. For instance Bergson (*op. cit.*) says "Nature is more and
better than a plan in course of realization. A plan is a term assigned to a
labor: it closes the future whose form it indicates. Before the evolution
of life, on the contrary, the portals of the future remain wide open. It
is a creation that goes on for ever..."

Although the relation between power and creation goes back to the
Old Testament, it appears that Nietzsche was the first philosopher to
note that the highest form of power is given in creation. For Nietzsche
(1888/1968), "The world is a work of art that gives birth to itself."

The evidence that there is a high degree of contingency in the Universe, which nowadays is difficult to question in the light of the quantum
phenomena and the biological evolution, would therefore result in a creative process in which living beings and mankind play a key role as
generators of novelty in the world. Paraphrasing Einstein, Karl Popper
said: "Appealing to his way [Einstein's] of expressing himself in theological terms, I said: If God had wanted to put everything into the
world from the beginning, He would have created a universe without

change, without organisms and evolution, and without man and man's experience of change. But He seems to have thought that a live universe with events unexpected even by Himself would be more interesting than a dead one."

According to Greek mythology humans and animals were created by the Titans: Zeus asked his sons Prometheus and Epimetheus to come to Earth and create humans and animals giving each of them a gift. Prometheus got to work making humans in the image of the gods whereas Epimetheus worked on the animals and gave each of them a gift. When finally Prometheus finished his work he asked Epimetheus to show him the gifts in order to choose one for humans. But Epimetheus confessed ashamed that he had exhausted all his gifts. That is how humans were created incomplete. In humans the fragmentary nature of each individual becomes self-conscious. Humans are a permanently unsatisfied fragment under their skin. They are constantly trying to achieve a higher state of existence or to be other and to transcend the everyday existence. In that search they fall into the greatest nonsensical pursuits but also the most sublime creations in science, art, moral or religion.

18.2 God

We adopt as we have already suggested a naturalistic conception of God as the state of the Universe in the framework of a monism of potentialities differs in numerous aspects of the traditional one. A set of questions arises that, given the state of advancement of our current knowledge, admit more than one possible option in their answers. Perhaps we will never know which is the correct one. Each option lays out a specific set of questions that have to be analyzed in order to get a complete conception and are a demonstration of how enriching the naturalist position can be when one reflects about this issue. We will analyze the options that appear most plausible from the strictly naturalist point of view, paying attention to the state of the art of our current scientific knowledge. We will not speculate which of the alternatives fits better our prejudices about the divine nature. The conclusion can be surprising and far away from the traditional theist positions. It can be considered as an exercise that only can reveal glimpses of a complete and coherent conception stemming from facts of which we only have partial knowledge.

18.2.1   Finite or infinite Universe

With the growing confirmation of the ideas about the Universe undergoing an inflationary period, in particular in the scenario of eternal
inflation as we discussed in chapter 12 and the preliminary results of
cosmologies based on quantum gravity the idea of an eternal and spatially infinite Universe has consolidated. The Universe contains regions
where inflation continues eternally and regions like our observable universe where the expansion has stopped being inflationary. The possibility of a very large but finite in space and time Universe cannot be
completely ruled out. We will adopt the most plausible scenario: the
first one, an infinite Universe.

18.2.2   What do we mean by the state of the Uni-
verse?

It depends on the interpretation of quantum mechanics considered. Some,
like the Modal or the Many Worlds interpretations assume that the state
of the Universe always evolves obeying Schrödinger's equation. Equiva-
lently, they assume that the state of the Universe is given once and for
all. One could call them eternal state interpretations. Others, which
include the Copenhagen or textbook interpretation and the Ghirardi–
Rimini–Weber interpretation assume that the state changes every time
an event takes place. One could call them the collapsing state interpretations. Once again, we face two alternatives with their critics and
defenders. The fact that two alternatives so different can coexist among
the experts show how plagued with subtleties is the process of producing
events (or measurements) and explains the difficulties one encounters in
finding a totally satisfactory solution.

An alternative that puts forward an intermediate point of view is
the Montevideo Interpretation. As we discussed, it is particularly appropriate to establish an ontology of objects and events. Let us recall
that basically this interpretation assumes that in order to realize predictions about the behavior of the physical systems one needs to use real
clocks subject to the rules of quantum mechanics. Schrödinger's evolution —called unitary— that is postulated by quantum mechanics is only
rigorously valid when the evolution is described in terms of an ideal time
not subject to the quantum laws. For local observers in the Universe,
like us, which necessarily use clocks ruled by quantum mechanics, the

evolution registered by our clocks departs from unitarity and leads to a loss of quantum coherence and the production of events with abrupt changes (collapse) of the states. This allows to present the process from two different points of view. From the point of view external to the observers the evolution take place in an ideal time and the state has a Schrödinger evolution; the eternal state picture is valid. We can say that the state of God remains eternally unchanged. For the observers of the world, change dominates, the events of the observed system are correlated wit the events of the clock. For different positions of the clock hands we have different configurations of the observed system. In order to realize predictions about natural objects it is convenient to use states that change each time an event takes place. The phenomenon of environmental decoherence ensures that the two descriptions, although apparently very different, are virtually indistinguishable for individual macroscopic fragments. The Montevideo Interpretation admits the co-existence of the two previous versions. This is because it assumes that events occur when one takes into account that in order to compute the evolution of the state of certain system we need to use real clocks. And also because events occur when the state of the system becomes indistinguishable from the one resulting from a collapse which is usually known as a proper statistical mixture

From the ontological point of view God is an individual object, that is a system —the Universe— in a given state, given once and for all. On the other hand we saw when we discussed the process of individuation by giving the example of a crystal, that the states of the latter change continuously when new events in which the crystal takes part take place. The crystal was not in this sense an "individual object" but a society of individual objects organized in time. Natural objects have an essential temporality and in analogy with the *actual occasions of experience* of Whitehead, they are essentially ephemeral. God and the world are therefore radically different in terms of their temporality.

18.2.3   Eternity and temporality

God and the natural world are from this point of view two completely different realities. As we have already seen, while God is an eternal individual object, that is a system in a state that is not subject to changes. The worlds' individual objects change each time an event occurs. The worlds' contingence is only limited by the potentialities present in God's state. Events are shared by both natures, for the world the evolution of certain events in terms of other determines the temporality while for God his necessary aspects do not preclude sharing with the world contingent aspects as the events, provided they do not conflict with the necessary ones. Without keeping the record of events the alternatives of choice for each individual object would be indefinite because in order to belong to the same individual they should share they history. In other words, events in our past, be they known or not by us, are absolutely definitive. They condition the present situation, in particular the states of the individual fragments. But those events, that have built up the current Universe, are not the current Universe, that only includes one event in each causally connected line of events. Their definitive nature, like the state of the Universe, only can exist in God. In fact, the exact disposition to produce events in systems laying in a certain spacetime region not only depends of the state of the Universe. It also depends on all the events that do not belong to the future of this region, and therefore one could say that they depend on the necessary and contingent aspects of God. Again, Whitehead's intuition seems to advance this possibility when he states that by the record of all achieved facts, the world achieves an "objective immortality" in God.

To assume that the nature of God is somehow dependent of contingent facts through the registration of events is not exempt of problems. As we saw when we studied general relativity there does not exist a present where events organize simultaneously side by side. Events evolve with respect to each other in a space-time that organizes them according to their ability to act on each other. There does not exist a unique notion of present. Basically any set of events without causal connection can be considered co-present. It does not result at all clear, when one speaks of the contingency derived from God's awareness of the development in the world, if it would be associated with any notion of time, since with current physical and cosmological ideas it would probably result meaningless. This is one of the many problems that need to be understood better and is related to the definition of time in the inflationary Universe.

The emergent vision is therefore one of a reciprocal relationship between God and Universe. God acts on the Universe in contributing to determine the disposition to act of its inhabitants. The Universe acts on God through the events that take place in it. God stops being an entity that acts on the world without being influenced by it. It could be said that this extends to the philosophical plane Einstein's belief when he rejected the existence of an absolute space since it makes no sense for something to act on another thing without being affected by it.

18.2.4   God and the Universe are Increate

We have analyzed extensively the surprising fine tuning for life of the laws of physics and the values of the fundamental constants. In the framework of a naturalist conception like we present here, with a Universe that can extend indefinitely in space and time and a conception that identifies God with the state of the Universe, it appears unacceptable to adopt the point of view that the physical laws are a divine creation. It would appear that God and the natural world are part of an increated reality whose fundamental laws, in what respects to the organization of the Universe and its quantum nature, have always been the same. To understand why these laws are suitable to harbor life and thought remits us to our regularist vision where the laws are regularities of entities whose intrinsic nature transcends them. If we recall that we have discovered that the world has a simultaneously material and mental nature, it does not appear absurd that its development would lead to increasing degrees of integration and manifestation of that double nature. In particular when one considers thinking living beings. Another point that approaches an answer to this question arises from recalling that God as dispositional state of the Universe is eternal and the same from the first to the last instant. Therefore all developments that the evolution of the Universe has generated through natural processes without a manifest teleology, potentially existed forever in God.

18.2.5   Natural theism

Natural theism like process theism (Hartshorne 1984) changes many of the traditional theist ideas in at least four aspects: a) The notion of divine perfection. b) The notion of an omniscient God, that is, that it knows it all. c) The notion of omnipotence and associated problem of evil. d) The appreciation of the material world.

In what concerns the notion of perfection, traditional theism conceives God as unsurpassable, it is that about which "nothing greater (or better) can be conceived." This notion is associated to the low esteem in which traditional theism has the world and its finite beings. In the traditional vision, God would not have stopped being perfect if he had not created the world. In the conception we are discussing the world and God are increate and both have a temporal dimension since in a world where freedom exists, future acts cannot be predetermined nor known beforehand. God is perfect in that it establishes eternally the slate of actions that are just and appropriate ethically and aesthetically for the situation in question. The concrete actions of the finite beings, be them good or evil, are known by God in that they produce event, but are not product of God's decisions and therefore God changes in time. But in fact for the natural theism —and against traditional theism— God does not make any decision. Only the finite beings, endowed with freedom, decide. Our actions in the world contribute or obstruct its full development towards which God, as a state of the universe, persuades us. Thinking our actions in aesthetic terms "creativity is inexhaustible and no actual creation can render further creation superfluous" (Hartshorne 1984). *To the innate perfection of the eternal state of the Universe and it potentialities we must add the open possibility of making them concrete in the world.*

God is not omniscient, that is, does not know everything. To begin with because God does not know the future of an event like it knows past events. Otherwise neither freedom nor creativity would make sense. The value of the world as a constantly ongoing contingent process in which what exists potentially becomes an act. This process in which God orients and participates would stop being possible. If the world and God have a temporal dimension the knowledge of future events can only be a possibility. But there exists a second limitation to divine knowledge that arises from the conception we are developing. The actions of an individual object, let us say ourselves, are determined by its dispositional state that manifests internally as passion or desire, to the extent that the eternal state of the Universe does not coincide with the state of the individual fragments. God cannot know or suffer in the same way that we do our passions and more in general our states. This lack of knowledge of our states is precisely the basis of the distinction between world and God. Indeed, in a Universe in which states and events always have a double aspect, one of which is the mental, it would not be possible to know the states of the individual without their nature being completely identical. Thus, while individual fragments and God share the knowledge of certain facts, the emotional tint that accompanies them being associated to their respective states is totally different. Paraphrasing

Spinoza, we could say that beatitude consists in perceiving the world from the perspective of God.

God is not omnipotent. Omnipotence has been understood as having the power of determining all event that occurs in the world. Implicit in the traditional theist way of understanding omnipotence is the ability to decide to act in a way or another. Hartshorne compares this way of understanding omnipotence with the tyrannical ideal of power. Understood in this way, the notion of omnipotence brings along the unsolvable problem of evil. Why does God allow an innocent to suffer. Why, if everything that happens is in the hands of God, it allows that evil take place and abstains from acting in the face of the greatest atrocities while being able to do so? In the natural theist conception we are presenting God only determines a set of dispositions to action that the subjects can freely decide to follow or not. The subjectivity of each individual stems completely from their state. God manifest itself only in each free act as an alternative associated to feeling of responsibility. Whitehead puts it like this: "God power is the worship he inspires." But maybe the statement falls short because many times this power does not seem to stem from a feeling of reverence. Many times the object of our responsibility is as perishable as we are. Facing it we are responsible not because it is a superior being but because it is a being that has as much right to live as we do. That being that moves us with the poverty of its contingent and perishable existence is the one who has the power to awake our responsibility and makes us act in his benefit. The feeling that predominates if at some point we feel facing a superior reality due to its plenitude or eternity is very different. It is not of responsibility but of reverence. But in both cases we are called to respond with the act in service of another. *God and world are two coexisting realities that form part of the process of self-creation of the Universe.*

On the other hand, notice that the events are always the product of acts of individual objects of the world; creation takes place in the world. In the monist perspective of potentialities, God or the state of the Universe only establish dispositions to action and there does not exist any form of supernatural intervention. It therefore establishes the impossibility of a divine being who can interrupt the world's development.

18.2.6   Mortality

Where in the history of life evolution can be draw a line between beings
with or without inwardness? Judging from what we observe and from
the principles of Darwinism some form of appetition is present in all the
living beings, at least in the animal kingdom. The same may be said
about the drive for survival and from this evidence and the observed
external behaviors induce that some form of fear of suffering and death
is also felt as the counterpart of the dispositional states of the organism.
In superior animals like vertebrates appetition reach new forms different from the vegetative ones with the appearance of motility and "the
interposition of distance between urge and attainment, i.e., in the possibility of a distant goal" (Jonas 2001). This new developments require
higher forms of perception and coordination, sentience and motility are
involved. Perceiving something as a goal and pursuing it requires the
desire of filling the gap between perception and fulfillment. The development of the mental pole achieves new heights with humans where, as
Prometeus myth tells us, its very existence becomes problematic and
what mankind must make of itself the subject of questioning.

The extensive and very rich process of development of sentient life
ruled by the forces of evolution and made possible by a Universe that
is hospitable to it in its laws and the very nature of the entities that
compose it presents the biggest paradox. The development of beings
capable of feeling and thinking, but, at the same time, capable of feeling with increasing intensity the contradiction between their impulse to
life and their destiny to die. The existentialists have insisted in how
finitude and perishability form essential part of the human being. But
the existentialism despises natural sciences and refuses to reflect about
what science reveals us about this condition as Camus already told us.
In this process of denial there is a double forgetfulness: that of value
and significance of the natural laws on one hand, and the central value
of the present moment, cornerstone of the construction of a scientific
knowledge in terms of events. For that reason for existentialism (for
example Heidegger) mankind is only a project and the existence is centered in the future, but a future that ultimately promises nothing, just
death.

What we have developed up to now of a natural theism gives some
hints of an answer to this paradox brought about by mortality. Let us
recall Whitehead's doctrine of the "objective immortality of the past".

In physical terms the present results from the confluence of the eternal state of the Universe with the sum of past events. This confluence would take place simultaneously in God as we have already proposed and the *individuals* that in that moment act. Thus, the immortality of past events in the mind of God would be accompanied by the "immortality of deeds". Those are the acts through which we contribute to the construction of that past. This intuition, already present in the origins of the Jewish religion and persists till today. That Jewish philosopher we have quoted so extensively in this book, Hans Jonas, says: "Might it not even be, to venture yet a step further, that what we thus add to the record is of surpassing import — not indeed for a future destiny of ours, but for the concern of the spiritual account itself kept by the unified memory of things? and that, although we mortal agents have no further stake in the immortality which our acts go to join, these acts of ours, and what through them we make of our lives, may just be the stake which an undetermined and vulnerable eternity has in us? And with our freedom, what precarious stake! —Are we then perhaps an experiment of eternity? our very mortality —a venture of the immortal ground with itself? our freedom— the summit of the ventures chance and risk?" The myth of Jonas suggests that God in the moment of creation put all of his being in play and it is up to the world, especially to mankind, endowed with liberty, to return to God the knowledge of its hidden essence.

From the conception outlined here a somewhat different myth emerges. God keeps its eternal being —the eternal state of the Universe— to conduct lovingly the world without violating the freedom of mankind. God guides us putting in front of us possible routes for us, as inhabitants of the world, to contribute to its self-creation and full realization. Our works, the events of the world, remain immortal in God's bosom and with that every instant of our life is eternally present in God. But the emotional tinting of this fleeting present and of the eternal present would be absolutely different in individual objects and in God. In one case the feeling would come from our state as individual object of the world, in the other of the eternal state of the Universe. That other Jewish thinker, Baruch Spinoza, stated that such divine vision was accessible to mankind and that the greatest virtue and satisfaction of the mind —its blessedness— consists in knowing the inmost essence of things and and their role in the wider scheme of beings. This "intellectual love of God, which arises [from this kind of knowledge] is eternal."

Chapter 19

Conclusions

The dominant naturalism has taken a particularly pernicious and uncritical shape resulting from an analysis of matter that still remains attached to the concept of particle. Although physicists understand well that when we refer to the fundamental components of the Universe we are talking about quantum fields and not particles, we still implicitly keep on thinking about them as particles. For instance, in the Wikipedia article for "Matter" it says: "All the objects from everyday life that we can bump into, touch or squeeze are composed of atoms. This atomic matter is in turn made up of interacting subatomic particles—usually a nucleus of protons and neutrons, and a cloud of orbiting electrons. Typically, science considers these composite particles matter because they have both rest mass and volume. By contrast, massless particles, such as photons, are not considered matter, because they have neither rest mass nor volume..." We therefore see that although the particles are just a possible behavior of a quantum field in certain particular states, we keep on referring to them as fundamental. We reserve for them the name of matter, which we deny to other equally important manifestations of the same quantum fields like light or the space-time of general relativity. Starting with that archaic notion of matter we wish to explain life and consciousness and get rid of their problematic aspects.

The understanding of a Universe basically inclined towards generating life and though requires to think of matter in more modern terms. Our proposal in this book has been to analyze it in terms of systems, states and events together with their properties. These are the basic concepts of a quantum Universe. We have verified that the resulting vision is at least more adequate to understand consciousness as a biological phenomenon. An advancement that helps this analysis was to understand that the notion of quantum state accounts for the emergence of composite systems with downward causation. This opens the possibility of understanding the effects of the mind on matter without reducing it to a mere epiphenomenon. The elaboration of these ideas, motivated by understanding the existence of a bio-friendly Universe has led us to propose some primary hypotheses that allow to set the foundations for human freedom. In addition, to develop a conception inspired in naturalism, more in tune with recent scientific results, and that radically departs from the traditional forms of materialism.

It is very possible that new developments in science will change radically the conclusions we reached. The search for a coherent and unified conception of the human experience that can explain how we and the other live beings form integral part of a world where our materiality and spirituality are not contradictory but complementary is a needed task that compels us not to take anything for granted. That is what we attempted to do in this book.

Bibliography

Aaronson, S. (2005) "Are Quantum States Exponentially Long Vectors?" arXiv.org:quant-ph/0507242.

Agaba, M., Ishengoma, E., Miller, W., McGrath, B., Hudson, C., Bedoya Reina, O., Ratan, A., Burhans, R., Chikhi, R., Medvedev, P., Praul, C., Wu-Cavener, L., Wood, B., Robertson, H., Penfold, L., Cavener, D. (2016) *Nature Communications* 7, 11519.

Alexander, S. (2010) "Space, time and deity: the Gifford lectures at Glasgow 1916-1918", reprinted by Nabu Press, Charleston, SC, USA.

Anastopoulos, C. (2008) "Particle or Wave: The Evolution of the Concept of Matter in Modern Physics", Princeton University Press, Princeton, NJ, USA.

Bassi, A. and Ghirardi, G. (2003) *Physics Reports* 379, 257.

Bedau, M. (2002) *Principia: an international journal of epistemology*, 6, 5.

Bedau, M. (2003) *Principia*, 6, 5.

Bell J S (1989) in Deser, S. and Finkelstein, R.J. "Themes in Contemporary Physics II", World Scientific, Singapore.

Bergson, H. (1907/2005) "Creative evolution" Cosimo Books, New York, NY, USA.

Berkeley, G. (2011) "A Treatise Concerning The Principles of Human Knowledge", CreateSpace, Seattle, WA, USA.

Bernstein, J. (2011) "Quantum leaps", Harvard University Press, Boston, MA, USA.

Bhaskar, R. (2008) "A realist theory of science", Routledge, New York, NY, USA.

Bohr, N. (1949) "Discussions with Einstein on Epistemological Problems in Atomic Physics" in Schilpp, p. (1949), "Albert Einstein: philosopher scientist" Cambridge University Press, Cambridge, UK.

Bohr, N. (1956) Letter to von Weiszäcker March 5th.

Bohr, N. (1995) "The Philosophical Writings of Niels Bohr, Vol. 3: Essays 1958-1962 on Atomic Physics and Human Knowledge" Ox Bow Press, Woodbridge, CT, USA.

Bohr, N. (2011) "Atomic Theory and the Description of Nature: Four Essays with an Introductory Survey", Cambridge University Press, Cambridge, UK.

Born, M. (2004) "The Born–Einstein letters 1916-1955" Palgrave Macmillan, New York, NY, USA.

Brentano, F. (1995) "Psychology from an Empirical Standpoint" Routledge, New York, NY, USA.

Broad, C. D. (2008) "The mind and its place in Nature (Philosophy of mind and language)", Routledge, New York.

Brickher, P. (2014) "Ontological Commitment", The Stanford Encyclopedia of Philosophy (Winter 2014 Edition), Edward N. Zalta (ed.), URL = ¡http://plato.stanford.edu/archives/win2014/entries/ontological-commitment/¿.

Brook, A (2010) "Unified consciousness and the self" in J. Shear and S. Gallaher "Models of the self", Imprint Academic, Charlottesville, VA, USA.

Brukner, C. and Kofler, L. (2010) arXiv.org:1009.2654 [quant-ph].

Buchler, J. (1955) "Philosophical writings of Peirce", Dover, New York, NY, USA.

Burke, C. (2012) "Confessions of St. Augustine" http://www.cormacburke.or.ke/node/1419

Busch, P. (2002), *Studies in the History and Philosophy of Modern Physics* 33B, 517-539.

Butterfield, J. (2011) *Foundations of Physics* 41, 920.

Butterfield, J. (2015) *Studies in the History and Philosophy of Modern Physics* A52, 75.

Camus, A. (1942/1955) "The myth of sisyphus", Penguin Random House, New York, NY, USA.

Carroll, J. (2011) "Laws of nature" in Edward N. Zalta (ed.) (2011) Stanford Encyclopedia of Philosophy.

Carter, B. (1974) "Large number coincidences and the anthropic principle in cosmology" In "Confrontation of cosmological theories with observational data; Proceedings of the Symposium, Krakow, Poland, September 10-12, 1973. (A75-21826 08-90)" Dordrecht Reidel Publishing Co., Dordrecht, Germany p. 291-298.

Casati, R. and Varzi, A. (2010) "Events", in Edward N. Zalta (ed.) (2010) The Stanford Encyclopedia of Philosophy.

Cassan, A. and Kubas, D. and Beaulieu, J.-P. and Dominik, M. and Horne, K. and Greenhill, J. and Wambsganss, J. and Menzies, J. and Williams, A. and Jörgensen, U. G. and Udalski, A. and Bennett, D. P. and Albrow, M. D. and Batista, V. and Brillant, S. and Caldwell, J. A. R. and Cole, A. and Coutures, C. and Cook, K. H. and Dieters, S. and Prester, D. D. and Donatowicz, J. and Fouqué, P. and Hill, K. and Kains, N. and Kane, S. and Marquette, J.-B. and Martin, R. and Pollard, K. R. and Sahu, K. C. and Vinter, C. and Warren, D. and Watson, B. and Zub, M. and Sumi, T. and Szymański, M. K. and Kubiak, M. and Poleski, R. and Soszynski, I. and Ulaczyk, K. and Pietrzyński, G. and Wyrzykowski, Ł., (2012) *Nature* 481, 167.

Chalmers (1989) "The First-Person and Third- Person Views (Part I)", http://consc.net/notes/first-third.html

Chalmers, D. (1996) "The Conscious Mind" Oxford University Press, Oxford, UK.

Chakravartty, A. (2007) "A metaphysics for scientific realism. Knowing the unknowable", Cambridge University Press, Cambridge, UK.

Clarke, S. (1717) "A Collection of Papers, which passed between the late Learned Mr. Leibnitz, and Dr. Clarke, In the Years 1715 and 1716", James Knapton, London, UK.

Clarke, D. M. (1644/2006) "The principles of philosophy" in "Descartes: A Biography" Cambridge University Press, Cambridge, UK.

Clifton, R. (2004) "Quantum entanglements. Selected papers", edited by Butterfield, J. and Halvorson, H., Oxford University Press, Oxford, UK.

Cohen-Tannoudji, C., Diu, B., Laloë, F. (2006) "Quantum mechanics" Wiley, New York, NY, USA.

Collini, E. , Wong, C., Wilk, K., Curmi, P., Brumer, P and Scholes G. (2010) *Nature*, 463, 644.

Conway Morris, S. (2004) "Life's solution" Cambridge University Press, Cambridge, UK.

Cooper, G. (2000) "The Cell, 2nd edition", Sinauer Associates, Inc., Sunderland, MA, USA.

Cottingham, C., Stoothoff, J. and Murdoch, D. (translators) (1985) "The philosophical writings of René Descartes" Cambridge University Press, Cambridge, UK.

Crane, T. (2001) "The Significance of Emergence" in B. Loewer and G. Gillett (eds) "Physicalism and Its Discontents" Cambridge University Press, Cambridge, UK.

Crane, T. (2010) "Cosmic Hermeneutics vs. Emergence: The Challenge of the Explanatory Gap" in "Emergence in Mind", Cynthia Macdonald and Graham Macdonald (eds.) Oxford University Press, New York, NY, USA.

Curley, E. (translator) (1985) "The collected works of Spinoza", Princeton University Press, Princeton, NJ, USA.

Dalton, J., Thomson, T. Wollaston, W. H. (1893/2011) "Foundations of atomic theory" Nabu Press, Charleston, SC, USA.

Damper, W. C. (1968) "A History of Science and its relations with Philosophy and Religion", Cambridge University Press, Cambridge, UK.

Damasio, A. (1989) "Concepts in the brain", *Mind & Language* 4, 24.

Darwin, C. (2010) "The descent of man", Dover, Mineola, NY, USA.

Darwin, C. (1859/1999) "The origin of species", Bantam Classics, New York, NY, USA.

Darwin, C. (1862/1977) "Fertilisation of orchids" in Freeman, R. B. (1977) "The Works of Charles Darwin: An Annotated Bibliographical Handlist." 2nd edn. Dawson, Folkestone, UK.

Davidson, D. (1970), "Mental events", reprinted in Davidson, D. (2001) "Essays on actions and events" Oxford University Press, New York, NY, USA.

Davies, P. (2006) "The Goldilocks Enigma: Why Is the Universe Just Right for Life?" Allen Lane, London, UK.

Dawkins, R. (1976/2006) "The selfish gene" Oxford University Press, Oxford, UK.

Dawkins, R. (2003) "A Devil's Chaplain" Houghton Mifflin, Boston, MA, USA.

Della Rocca, M. (2008) "Spinoza," Routledge, New York, NY, USA.

Derrida, J. (2001) in "D'Ailleurs", documentary by Safaa Fathy, http://icarusfilms.com/new2001/derr.html http://www.youtube.com/watch?v=rhUy98VOEO4&feature=share

Descartes, R. (1644/2010) "The Principles Of Philosophy," Kessinger Publishing LLC, Whitefish, MT, USA.

de Sousa, R. (1987) "The Rationality of Emotion", MIT Press, Cambridge, MA, USA.

d'Espagnat, B. (1990) *Foundations of Physics* 20, 1147.

d'Espagnat, B. (1995) "Veiled reality", Addison Wesley, New York.

d'Espagnat, B. (2006) "On physics and philosophy", Princeton University Press, Princeton, NJ, USA.

Deutsch, D. (1999) *Proceedings of the Royal Society of London*, A455, 3129.

Drake, S. (translator) (1962) Galileo Galilei's "Dialogue Concerning the Two Chief World Systems, Ptolemaic and Copernican, Second Revised edition", University of California Press, Berkeley, CA, USA.

de Wall, F (1997), "Bonobo the forgotten ape", University of California Press, Berkeley, CA, USA.

de Witt, B. and Graham, N. (1973) "The many worlds interpretation of quantum mechanics", Princeton University Press, Princeton, NJ, USA.

Dieks, D. (1989) "Quantum mechanics without the projection postulate and its realistic interpretation" *Found. Phys.* 19, 1397.

Dirac, P. A. M. (1982) "The principles of quantum mechanics", Oxford University Press, Oxford, UK.

Dorato, M. (2015), *Topoi*, 34, 369.

Einstein, A. (1917). Letter from Einstein to Schlick, 21 May 1917, Einstein Archives 21-618.

Einstein, A. (1918) Address to the Berlin Physical Society, reprinted in Einstein, A (1954) "Ideas and Opinions", Crown, New York, NY, USA.

Ellis, G. (2016), "How can physics underlie the mind?: Top-down causation in the human context (The Frontiers Collection)", Springer, Berlin, Germany.

Everett III, H. (1957) *Reviews of Modern Physics*, 29, 454.

Faye, J. (2014) "Copenhagen Interpretation of Quantum Mechanics" The Stanford Encyclopedia of Philosophy (Fall 2014 Edition), Edward N. Zalta (ed.), http://plato.stanford.edu/archives/fall2014/entries/qm-cop enhagen .

Feschbach, N. (1975), *The Counseling Psychologist* 5, 25.

Feyerabend, P. (2010) "Against method" Verso, New York, NY, USA.

Fiske, J. (1874) "Outline of Cosmic Philosophy", as quoted by James, W. (1950) " Principles of psychology, vol 2" Dover, Mineola, NY, USA.

French, S. (1989) *Philosophical Studies: An International Journal for Philosophy in the Analytic Tradition*, 55, 1.

Frenkel, A. (2010) "A Review of Derivations of the Space-Time Foam Formulas," arXiv:1011.1833 [quant-ph].

Gambini, R., Porto, R. (2001) *Physical Review* D 63, 105014

Gambini, R., Porto, R., Pullin, J., Torterolo, S. (2009) *Phys. Rev.* D 79, 041501.

Gambini, R., García-Pintos, L. P., Pullin, J (2010/2011) *International Journal of Modern Physics D* 20, 909 [arXiv:0905.4222 [quant-ph]]; *Foundations of Physics* 40, 93 [arXiv:1009.3817 [quant-ph]]; Gambini, R., Pullin, J., (2015) arXiv:1502.03410 [quant-ph].

Gambini, R., Lewowicz, L., Pullin, J. (2015) *Foundations of Chemistry* 17 pp 117-127.

Gambini, R., Pullin, J. (2016) *International Journal of Quantum Foundations* 2, 89-108.

Gadamer, H.-G. (1981) "Reason in the age of science", MIT Press, Boston, MA, USA.

García, D., Pallardó, C., Ríos Insua, D., Moreno, R. and Redchuk, A. (2011) *ERCIM News* 84, 19.

Ghirardi, G. (2007) "Sneaking a Look at God's Cards, Revised Edition: Unraveling the Mysteries of Quantum Mechanics", Princeton University Press, Princeton, NJ, USA.

Hacking, I (1983) "Representing and Intervening: Introductory Topics in the Philosophy of Natural Science" Cambridge University Press, Cambridge, MA.

Hafele, J and Keating, R (1972) *Science* 177, 4044.

Hartshorne, C. (1979) "Whitehead's revolutionary concept of prehension" *Internatinoal Philosophical Quarterly* 19, 256-276.

Hartshorne, C. (1984) "Omnipotence and other theological mistakes", State University of New York Press, Albany, NY, USA.

Healey, R. (1989) "The philosophy of quantum mechanics: an interactive interpretation", Cambridge University Press, Cambridge, UK.

Healey, R. (1999) "Holism and nonseparability in physics" in Stanford Encyclopedia of Philosophy http://plato.stanford.edu

Healey, R. (2007) "Gauging what's real", Oxford University Press, New York, NY, USA.

Heidegger, M. (1936/1979), Krell, D. (translator) "Nietzsche: The will to power as art" Harper and Row, New York, NY, USA.

Heisenberg, W. (1958/2007) "Physics and Philosophy: The Revolution in Modern Science" HarperCollins, New York, NY, USA.

Helmholtz, H. (1853) " On the conservation of force" in "Scientific Memoirs, Natural Philosophy" J. Tyndall (translator) William Francis, London, UK.

Helmholtz, H. (1867) "Handbuch der physiologischen Optik" Leopold Voss, Leipzig, available online at http://www.archive.org/details/handbuchderphysi00helm

Hobbes, T. (1982) "Leviathan", Penguin Classics, Penguin, Edinburgh, UK.

Hoffmeyer, J. (2010) "Semiotic Freedom: An Emerging Force" in Niels Henrik Gregersen and Paul Davis (eds.), "Information and the Nature of Reality: From Physics to Metaphysics," Cambridge University Press, Cambrige, UK 185-204.

Hollands, S. and Wald, R. (2004) *General Relativity and Gravitation* 36, 2595.

Honderich, T. (1990) "Mind and brain: a theory of determinism, volume 1" Oxford University Press, New York, NY, USA.

Howard, D. "Einstein's Philosophy of Science", The Stanford Encyclopedia of Philosophy (Summer 2010 Edition), Edward N. Zalta (ed.)

Huang, Z., Wang, H. and Kais, S. (2006) *Journal of Modern Optics* 53, 10.

Hume, D. (1999) "An Enquiry concerning Human Understanding", Oxford University Press, Oxford, UK.

Humphreys, P. (1997) *Philosophy of science*, 64, 1.

Jackson, F. and Ph. Pettit (1992), "Structural Explanation in Social Theory", in Charles, D. and Lennon, K. (eds.), "Reduction, Explanation, and Realism" Oxford, Clarendon, pp. 97-13

Huxley, T. H. (1893/2001) "Collected Essays: Volume 1. Method and Results" Adamant Media Corporation, Boston, MA, USA.

Huygens, C. (2010) "Oeuvres Completes de Christiaan Huygens", Nabu Press, Charleston, SC, USA.

Israel, J. (2002) "Radical Enlightenment: Philosophy and the Making of Modernity 1650-1750" Oxford University Press, Oxford, UK.

Jammer, M. (1974) "The Philosophy of Quantum Mechanics: The Interpretations of Quantum Mechanics in Historical Perspective" Wiley, New York, NY, USA.

Johansson, I. (2006) "Ontological Investigations: An Inquiry into the Categories of Nature, Man, and Society" Ontos Verlag, Heusenstamm, Germany.

Jonas, H. (2001) "The phenomenon of life: toward a philosophical biology", Northwestern University Press, Chicago, IL.

Kant, I. (2010) "Groundwork of the Metaphysics of Morals" readaclassic.com, Cedar Lake, MI, USA.

Károlhyázy F., Frenkel A., Lukács, B. (1986). "Gravity in the reduction of the wavefunction", in R. Penrose and C. Isham, (eds.), "Quantum concepts in space and time". Oxford University Press, Oxford, UK.

Kauffman, S. and Clayton, P. (2006) *Biology and Philosophy*, 21, 4.

Keefe, A. and Szostak, J. (2001) *Nature*, 410, 715.

Kim, J. (1998) "Mind in a Physical World: An Essay on the Mind-Body problem and Mental Causation. "Representation and Mind Series". MIT Press, Cambridge, MA, USA.

Kim, J. (2008) "Making sense of emergence" in Bedau, M. and Humphreys, P. (eds) "Emergence. Contemporary readings" MIT Press, Cambridge, MA, USA p 127.

Klever, W. (1995) "Spinoza's life and works" in Garrett, D. "The Cambridge Companion to Spinoza", Cambridge University Press, Cambridge, UK.

Kochen, S. (1985) "A new interpretation of quantum mechanics" in Mittelstaedt, P. and Lahti, P. (eds) "Symposium on the foundations of modern physics" , World Scientific, Singapore.

Kochen, S. and Specker, E. (1967) *Journal of Mathematics and Mechanics* 17, 59.

Laplace, P. S. de (1814/2010) "A Philosophical essay on Probabilities", Nabu Press, Charleston, SC, USA.

Laue, M. (1913) "Das Relativitätsprinzip" in *Jarhbücher der Philosophie* S 1, p99, quoted in Cassirer, E. (1980) "Substance and function and Einstein's theory of relativity", Dover, Mineola, NY, USA.

Lewis, D. (1986) "Philosophical Papers", Volume II, Oxford: Oxford University Press.

Linde, A. (1982) *Physics. Letters* B108, 389.

Linde, A. (1986) *Physics Letters* B175, 395.

Lloyd, S. (2010) "The computational universe" in Paul Davies and Neils Gregersen "Information of the Nature of Reality," Cambridge University Press, Cambridge, UK.

Lucas, Jean Maximilien (1899) as quoted in Garrett, D. (1996) "The Cambridge companion to Spinoza", Cambridge University Press, Cambridge, UK.

Luisi, P. L. (2016) "The emergence of life" Cambridge University Press, Cambridge, UK.

Lynch, M. (2007) "The origins of genome architecture" Sinauer Associates, Inc., Sunderland, MA, USA.

Margulis, L. and Sagan, D. (1997) "Slanted truths: essays on Gaia, symbiosis and evolution", Springer, Berlin, Germany.

Maudlin, T., (1998) "Parts and whole in quantum mechanics" in E. Castellani (ed.) "Interpreting bodies", Princeton University Press, Princeton, NJ, USA.

Maudlin, T. (2007) "The metaphysics within physics", Oxford University Press, Oxford, UK.

Maturana, H., Varela, F. (1998) "The Tree of Knowledge: The biological roots of human understanding" (Revised Edition) Shambhala Publications, Boston, MA, USA.

Mavromatos, N. (2011) *Journal of Physics Conference Series*, 305, 012008.

Maxwell, G. (1962) "The ontological status of theoretical entities" in Feigl, H. and Maxwell, G. "Scientific explanation, space and time, vol. 3 Minnesota Studies in the Philosophy of Science" University of Minnesota Press, Minneapolis, MN, USA 3-15.

Mayr, E. (1998) "This is biology: the science of the living world" Belknap Press, Cambridge, MA, USA.

McFadden, J., Al-Khalili, J. (2016) "Life on the edge", Broadway Books, New York, NY, USA.

McLaughlin, Brian (1992) "The Rise and Fall of British Emergentism" in Beckermann, A., Flohr, H., and Kim, J., eds. (1992). Emergence or Reduction? Berlin: Walter de Gruyter.

Mermin, D. (1985) "Is the moon there when nobody looks? Reality and the quantum theory", *Physics Today*, April, p38.

Michelson, A. and Morley, E. (1887), *American Journal of Science* 34, 333.

Mill, J. S. (1843) "A system of logic: ratiocinative and inductive", reprinted by University Press of the Pacific (2002), Honolulu, HI, USA.

Molina Y Vedia, C., Grimm, R. (1995) "Philosophical writings" Bloomsbury Academic, London, UK.

Monegal, R. and Reid, A. (1981) "Borges: A Reader: A selection from the writings of Jorge Luis Borges" Plume/Dutton, New York, NY, USA.

Monod, J. (1970/71) "Chance and necessity: An essay on the natural philosophy of modern biology", Alfred A. Knopf, New York, NY, USA.

Monroe, K. (1996) "The heart of altruism. Perceptions of a common humanity", Princeton University Press, Princeton, NJ, USA.

Monroe, K. (2002) "Explicating altruism" in Post, S., Underwood, L., Schloss, J. and Hurlbut, W. (2002) "Altruism & Altruistic Love: Science, Philosophy & Religion in Dialogue", Oxford University Press, Oxford, UK.

Moss, L. (2004) "What genes can't do" (Basic bioethics) MIT Press, Cambridge, MA, USA.

Myrvold, W. (2003) *International studies in the philosophy of science* 17, 7.

Nagel, E. (1969) "Principles of the theory of probability" University of Chicago Press, Chicago, IL, USA.

Nagel, E. (1979) "The Structure of Science", Hackett, New York, NY, USA.

Nagel, T. (1974) *Philosophical Review*, pp. 435-50.

Nagel, T. (1994) in Warner, R. and Szubka, T. (1994) "The mind-body problem: a guide to the current debate" Wiley–Blackwell, New York, NY, USA.

Ng, Y. and van Dam, H. (1995). *Annals of the New York Academy of Sciences* 755, 579.

Nietzsche, F. (1887/2009) "On the genealogy of morals", Oxford University Press, Oxford, UK.

Nietzsche, F. (1888/1968) "The will to power" Vintage, New York, NY, USA.

Nishitani, K. (1990) "The self-overcoming of nihilism" Parks, G. translator. State University of New York Press, Albany, NY.

Niven, W. D. (2010) "The scientific papers of James Clerk Maxwell" Nabu press, Charleston, SC, USA.

Northrop, F. (1949) "Einstein's conception of science" in Schilpp, p. (1949), "Albert Einstein: philosopher scientist" Cambridge University Press, Cambridge, UK.

Oberhummer, H., Csótó, A., Schlattl, H. (2000) Science 289, 88.

O'Loan, O. J. and Evans, M. R. (1999), *Journal of Physics A* 32, L99.

Omnés, R. (1994) "The interpretation of quantum mechanics", Princeton University Press, Princeton, NJ, USA p. 98.

Owen, H (1971) *Religious Studies*, 7, 175.

Papineau, David, "Naturalism", The Stanford Encyclopedia of Philosophy (Fall 2015 Edition), Edward N. Zalta (ed.), URL = ¡http://plato.stanford.edu/archives/fall2015/entries/naturalism/¿.

Passmore, J. (1967). "Logical Positivism." In Edwards, P. (Ed.). The Encyclopedia of Philosophy (Vol. 5, 52-57). Macmillan, New York, NY, USA.

Penrose, R. and Hameroff S. R. (1995) *Journal of Consciousness Studies* 2, 99;

Peres, A. (1984) *Found. Phys* 14, 1131.

Pink, T. (2004) "Free will. A very short introduction" Oxford University Press, Oxford, UK.

Plankar, M., Jerman, I. and Krašovec, R. (2011) *Progress in Biophysics and Molecular Biology* 106, 380.

Preston, S. and de Waal F. (2002) "The communication of emotions and the possibility of empathy in animals." In "Altruistic love: Science, philosophy, and religion in dialogue" ed. S. Post, L. G. Underwood, J.

P. Schloss, & W. B. Hurlburt; Oxford University Press, Oxford, UK pp. 284-308.

Priest, S. (1992) "Theories of the Mind: A Compelling Investigation into the Ideas of Leading Philosophers on the Nature of the Mind and Its Relationship to the Body" Mariner books, Boston, MA.

Putnam, H. (1962) "What Theories Are Not?" in Nagel, E., Suppes, P. and Tarski, A. (eds.) (1962) "Logic, Methodology and Philosophy of Science: Proceedings of the 1960 International Congress" Stanford University Press, Palo Alto, CA, USA.

Radman, M., Matic, I., Taddei, F. (1999) *Annals of the New York Academy of Sciences* 870, 146.

Ramachandran, V. S. and Hirstein, W. (1997) "Three laws of qualia", *Journal of Consciousness Studies* 4, 429-58.

Rescher, N. (1970) "Scientific explanation" Free Press, New York, NY, USA.

Ruddick, C. (1949) *Philosophy of Science*, 16, 89.

Ruse, M. (2002) "A Darwinian naturalist's perspective on altruism" in Garrard Post, S., Underwood, L. and Schloss, J. "Altruism and altruistic love: science, philosophy & religion in dialogue", Oxford University Press, Oxford, UK.

Russell, B. (1903/2012) "Mysticism and logic" CreateSpace, New York, NY, USA.

Russell, B. (1927/2007) "The analysis of matter", Spokesman books, Nottingham, UK.

Russell, B. (1921/2011) "The analysis of mind", CreateSpace, Seattle, WA, USA.

Russell, B. (1925/2004) "What I believe", Routledge, New York, NY, USA.

Sagan, C. (1995) *Skeptical Inquirer*, vol 19.1.

Sarkar, S and Pfeifer, J. (2006). "Physicalism: The causal impact argument". "The Philosophy of Science: N-Z, Index." Taylor & Francis, New York, NY, USA.

Sapp, J. (2003) "Genesis: the evolution of biology" Oxford University Press, Oxford, UK.

Salecker, H. and Wigner, E. P. (1958) *Physical Review* 109, 571.

Saunders, S (2010) in Saunders, S., Barrett, J., Kent, A. and Wallace, D. "Many Worlds?: Everett, Quantum Theory, and Reality ", Oxford University Press, Oxford, UK.

Schaffer, J. (2011) "Monism" in Stanford Encyclopedia of Philosophy http://plato.stanford.edu

Scheibe, E. (1973) "Logical Analysis of Quantum Mechanics (International series of monographs in natural philosophy)", Pergamon, London, UK.

Schilpp, A. (1998) "Albert Einstein, Philosopher-Scientist: The Library of Living Philosophers Volume VII" Open Court, Chicago, IL, USA.

Schlamminger, S., Choi, K.-Y., Wagner, T., Gundlach, J. and Adelberger, E. (2008). *Physical Review Letters*, 100,4.

Schlosshauer, M. (2010) "Decoherence: and the Quantum-To-Classical Transition (The Frontiers Collection)" Springer, Berlin, Germany.

Schopenhauer, A. (1839/1999) "Prize Essay on the Freedom of the Will", Cambridge University Press, Cambridge, UK.

Schroeder, T. (2004) "Three faces of desire" Oxford University Press, Oxford, UK.

Schrödinger, E. (1944/2012) "What is life. With Mind and Matter and Autobiographical Sketches (Canto Classics)" Cambridge University Press, Cambridge, UK.

Searle, J. (2007) "Biological naturalism" in Velmans, M., Schneider, S. (Editors) "The Blackwell companion to consciousness", Blackwell, New York, NY, USA.

Sellars, W. (1960/1963) "Philosophy and the scientific image of man". Reprinted in Sellars, W. (1963) "Empiricism and the philosophy of Mind", Routledge & Kegan Paul Ltd., London, UK.

Sheldrake, R. (2012) "Science delusion", Coronet, New York, NY, USA.

Shimony, A. (1978) *International Philosophical Quarterly* 18,3.

Smolin, L. (1999) "The life of the cosmos" Oxford University Press, Oxford, UK.

Sontag, S. (2002) "Styles of radical will" Picador, London, UK.

Sperry, R., (1987) *Journal of Mind and Behavior* 8, 37-66.

Spinoza, B. (1677/2005) "Ethics" Penguin Classics, Penguin, Edinburgh, UK.

Spinoza, B. (1663/2007) "Principles of Cartesian philosophy" Philosophical Library, New York, NY, USA.

Spinoza, B. (1670/2007) "Theological-Political treatise" Israel, J. (editor), Cambridge University Press, Cambridge, UK.

Stanford (2011) "The Stanford Encyclopedia of Philosophy" http://plato.stanford.edu

Stoljar, D. (2009) "Physicalism" in Stanford Encyclopedia of Philosophy http://plato.stanford.edu

Strawson, G. (2010) "The self" in J. Shear and S. Gallaher "Models of the self", Imprint Academic, Charlottesville, VA.

Stumpf, S. (2007) "Socrates to Sartre and Beyond" McGraw-Hill, New York, NY, USA p 357.

Swartz, N. (2003/1985) "The concept of physical law", second edition (2003) (available online at http://www.sfu.ca/philosophy/physical-law/). Original (1985), Cambridge University Press, Cambridge, UK.

Tegmark, M. (1993) *Foundations of Physics Letters*, 6, 571-590.

Tegmark, M. (2000) *Physical Review* E61, 4194.

Tegmark, M. (2008) *Foundations of Physics*, 38, 101.

Teller, P. (1986) *The British Journal for the Philosophy of Science*, 37, 71.

Tye, M. (2007) "Philosophical problems of consciousness" in Velmans, M., Schneider, S. (Editors) "The Blackwell companion to consciousness", Blackwell, New York.

van Fraassen, B. (1980) "The scientific image", Oxford University Press, Oxford, UK.

Vattay, G. Kauffman, S., Niiranen, Samuli (2012) "Quantum biology at the edge of quantum chaos", ArXiv:1202.6433.

Vedral, V. (2011) "Living in a quantum world" *Scientific American*, June.

Vilenkin, A. (1983) *Phys. Rev.* D27, 2848.

Viney, D., "Process Theism", The Stanford Encyclopedia of Philosophy (Spring 2014 Edition), Zalta, E. N. (ed.), URL = ¡http://plato.stanford.edu/archives/spr2014/entries/process-theism/¿.

Voltaire, F. (1932) "The Ignorant Philosopher" In "The Best Known Works of Voltaire", Blue Ribbon Books, London, UK.

von Neumann, J. (1996) "Mathematical Foundations of Quantum Mechanics", Princeton University Press, Princeton, NJ, USA.

von Weiszäcker, C.F. (1955) *Die Naturwissenschaften* 41, 521.

Wagner, A. (2015) "Arrival of the fittest: how Nature innovates" Penguin, New York, NY, USA.

Wandschneider, D. (2005), "Darwinism and philosophy", Notre Dame Press, Notre Dame, IN, USA.

Weinberg, S. (1974) *Scientific American* 231, 50.

Whitehead, A. N. (1925/1997) "Science and the modern world" Free Press, New York, NY, USA.

Whitehead, A. N. (1929/1979) "Process and reality" Free Press, New York, NY, USA.

Wildman, W. (2009) "Science and Religious Anthropology: A Spiritually Evocative Naturalist Interpretation of Human Life (Ashgate Science and Religion)" Routledge, New York, NY, USA.

Wilson, E. O. (1999) "The diversity of life", W. W. Norton, New York, NY, USA.

Wimsatt, W. C. (2007) "Aggregativity: Reductive heuristics for finding emergence", in M. Bedau, P. Humphreys (eds) "Emergence: contemporary readings in philosophy and science", MIT Press, Boston, MA, USA.

Wimsatt, W. C. (1998) *Philosophy of Science* 65, 267.

Wolf, A. (2010) "Spinoza's Short treatise on God, man and his wellbeing," Nabu Press, Charleston, SC, USA. http://www.yesselman.com/WolfIntroduction.htm

Wordsworth, W. (1802) "The world is too much with us" Wikipedia.

Zhu, J. (2004) *Canadian Journal of Philosophy* 34, 175-194.

Zurek, W. (1982) *Physical Review D* 26,1862.

Zurek, W. (1993) *Physics Today* 46, 81.

Zurek, W. (2003) *Review of Modern Physics* 75, 715.

Index